A History of
Gold Dredging
in Idaho

MINING THE AMERICAN WEST

Duane A. Smith, Robert A. Trennert, and Liping Zhu, editors

Boomtown Blues: Colorado Oil Shale, Andrew Gulliford

From Redstone to Ludlow: John Cleveland Osgood's Struggle against the United Mine Workers of America, F. Darrell Munsell

Gambling on Ore: The Nature of Metal Mining in the United States, 1860–1910, Kent A. Curtis

Hard as the Rock Itself: Place and Identity in the American Mining Town, David Robertson

High Altitude Energy: A History of Fossil Fuels in Colorado, Lee Scamehorn

A History of Gold Dredging in Idaho, Clark C. Spence

Industrializing the Rockies: Growth, Competition, and Turmoil in the Coalfields of Colorado and Wyoming, David A. Wolff

The Mechanics of Optimism: Mining Companies, Technology, and the Hot Spring Gold Rush, Montana Territory, 1864–1868, Jeffrey J. Safford

Mercury and the Making of California: Mining, Landscape, and Race, 1840–1890, Andrew Scott Johnston

The Rise of the Silver Queen: Georgetown, Colorado, 1859–1896, Liston E. Leyendecker, Duane A. Smith, and Christine A. Bradley

Santa Rita del Cobre: A Copper Mining Community in New Mexico, Christopher J. Huggard and Terrence M. Humble

Silver Saga: The Story of Caribou, Colorado, Revised Edition, Duane A. Smith

Thomas F. Walsh: Progressive Businessman and Colorado Mining Tycoon, John Stewart

Yellowcake Towns: Uranium Mining Communities in the American West, Michael A. Amundson

A History of Gold Dredging in Idaho

Clark C. Spence

UNIVERSITY PRESS OF COLORADO
Boulder

Published by University Press of Colorado
5589 Arapahoe Avenue, Suite 206C
Boulder, Colorado 80303

Printed in the United States of America

The University Press of Colorado is a proud member of
Association of American University Presses.

The University Press of Colorado is a cooperative publishing enterprise supported, in part, by Adams State University, Colorado State University, Fort Lewis College, Metropolitan State University of Denver, Regis University, University of Colorado, University of Northern Colorado, Utah State University, and Western State Colorado University.

∞ This paper meets the requirements of the ANSI/NISO Z39.48–1992 (Permanence of Paper).

ISBN: 978-1-60732-474-4 (cloth)
ISBN: 978-1-60732-475-1 (e-book)

Library of Congress Cataloging-in-Publication Data
Spence, Clark C.
A history of gold dredging in Idaho / Clark C. Spence.
pages cm
Includes bibliographical references.
ISBN 978-1-60732-474-4 (cloth) — ISBN 978-1-60732-475-1 (ebook)
1. Gold mines and mining—Idaho—History. 2. Placer mining—Idaho—History. I. Title.
TN413.I2S64 2016
622'.342209796—dc23

2015033570

Cover photograph: the Boston and Idaho Company's fifteen-foot bucket dredge, courtesy, Idaho State Historical Society, Boise, ISHS 72-112.b.

For Judith Austin

Contents

Acknowledgments

I owe a debt of gratitude to the staffs of several major libraries, especially the University of Illinois, as well as the Huntington and the Bancroft Libraries in California, and to a number of individuals and organizations. Alan Virta of Special Collections, Boise State University, provided pertinent documents from the Paul Tracy papers, and Lorna Tracy, Paul's daughter, approved usage of the text from her home in England. Cindy Schacher, an archaeologist with the Nez Perce–Clearwater Forest Service, sent important letters and books on dredging. Edgar Rice Burroughs, Inc., a California concern, furnished three rare late-nineteenth-century dredge icons. Jamie Edmondson, spokesperson for the Idaho Gold Fields Historical Society, Smith Collection, came up with a hard-to-find photograph of a dragline dredge. As director of the Lemhi County Historical Society and Museum in Salmon, Hope Benedict went beyond the call of duty and was able to provide several very striking pictures. At the Idaho State Historical Society in Boise, the busy work staff was responsible for digitally copying a wide range of dredge-related photos, with Layce Johnson handling the orders and seeing them to their destination.

Judith Austin was a key figure in helping select these photos. Without her help, this book would not have been possible. Finally, hats off to the staff of the University Press of Colorado for their expertise, patience, and understanding. Specifically, Jessica d'Arbonne has been the ideal editor.

A History of
Gold Dredging
in Idaho

Introduction

Gold is where you find it, is an old mining saying. Whenever it is found, it appears in its natural form in two different ways, depending on the geologic environment. One form is in hard-rock lode deposits, deposits that may have had different origins but which remain locked within the original solid rock formations. Even if a lucky prospector located a so-called quartz mine, it was clear that development was for corporate, not individual, effort and might take years before it returned a profit. The work called for more and more capital for tunneling, timbering blasting ore, transporting it underground, and hoisting it to the surface. Moreover, gold in hard-rock ores was often difficult to remove—"refractory" in the language of the engineers. In the worst cases, it might be in combination with other elements, which meant more intricate and costly processes would be needed. These kinds of deposits have produced the bulk of the world's gold to date.

The second type is the secondary deposit, which occurs in several forms. Residual deposits are usually found where rocks have weathered and deteriorated, but without water. They have not yet been washed away, nor have

DOI: 10.5876/9781607324751.c000

they moved. Generally, the residual lies near the site of the lode. Another type, bench deposits, is created when gold reaches a waterway, which subsequently dries up or changes course, leaving an old streambed on higher ground that drains into valleys. Stream or alluvial deposits are much more common and often contain more precious metal. As such, they constitute collections of loose particles of gold that are freed from their enclosing lode deposits by erosion, disintegration, or decomposition and then slowly washed down slopes into streams that quickly winnow out the lighter material. Thus, heavy minerals, including gold, platinum, tin, and some gemstone, are concentrated near or even within bedrock in such a way as to constitute workable deposits.[1]

From ancient times, such placer deposits were easier to find and work than lode mines, and more than one technique would carry over into early gold mining in the American West. The panning of gold was the simplest way for the prospector to test for placer gold. The early Romans had used a pan of some sort for that purpose, and widespread usage continued through the ages. By the time Californians brought it to Idaho, the pan had become standardized in sheet iron, eighteen inches in diameter and four inches deep, with sides slanting outward at an angle of thirty-seven degrees.[2] The miner threw a few handfuls of gold-bearing soil or gravel into the pan, poured two or so inches of water over it, then swirled the contents in a sideways rotary motion, thus washing the lighter material over the side and leaving the heavier gold behind. It took ten to twelve minutes to wash a pan, and a man could do fifty of them in a good day, but it was strenuous work, and unless the ground was exceptionally rich, it was hardly profitable.[3] Fortunately for the early comers, gold was plentiful and easily panned in the early days of all of Idaho's high-country placers—Pierce, Florence, Elk City, Warren, Salmon City, Stanley Basin, Boise Basin, and on the South and Middle Forks of the Boise River.

The rocker or cradle was a considerable improvement over the pan. This was a rectangular box, set at a downward angle and mounted on a rocking mechanism like that of a rocking chair. At the top was a removable hopper with a mesh screen or perforated iron plate; at the bottom was a series of cleats or "riffles." The gravel was dumped into the top, followed by a bucket of water, after which the cradle was rocked by hand to agitate the mixture. Any rocks were caught by the screen or perforated plate; the smaller wasted exited at the lower end with the water, and the heavy gold fell into the bottom of the box and was caught in the riffles. The machine was easily transported and did not require a constant flow of water. One man could operate it, but a team of three or four made it much more efficient: one or two to dig, one to dump the paydirt and water into the hopper, and another to work the rocker handle with vigor.[4]

The machine was an advancement over the sluice, which in its most primitive form was simply a long ditch with its bottom cleated with rocks, gravel, and holes to act as riffle bars; more often it took the shape of a long tom. Built of wood, the latter was ten feet to twenty feet long, a foot-and-a-half deep, and tapered at one end so that a number of them might be fitted together to form a sluice sometimes several hundred feet long. The lower end of the long tom, called the riddle, replaced the hopper of the cradle, and a heavily perforated iron strained out the large debris while allowing water and small gravel to fall into a riffle box where the gold was captured. Widely used throughout the West, such equipment needed an available running stream of water, which was usually supplied by running a flume from a nearby river or creek.[5]

Another approach, hydraulic mining, added a more sophisticated modern twist to a Roman technique described by Pliny the Elder in the first century. The episode pictured by Pliny involved a violent stream of water used to undermine a hill in a gold mining operation: "Then all at once . . . the mountain cleaveth in sunder and making a long chink, falleth down with such a noise and crack that is beyond the conceit of man's understanding."[6] As devised by Connecticut Yankee Edward Matteson in California in 1853, hydraulicking has been likened to directing a fire hose against a sand pile. By dropping a stream from a small reservoir above the mining site through a crude rawhide or canvas hose, miners could shoot a powerful head of water under strong pressure from a pivoting nozzle (a monitor) and literally wash away the sides of a gravel bank, then run the gravel through a long sluice for recovery of its gold. Within a few years, improvements in hoses and monitors allowed even more pressure—so strong that more than one worker died when struck by their stream. In 1870 the monitor manufactured by the Hoskins factory and called the "Little Giant" lent its name to any monitor, which became a "giant." The historian of California's great hydraulicking controversy referred to Hoskins's "Little Giant" as "essentially a piece of artillery and reminiscent of the Civil War."[7] Available water and hilly terrain were prerequisites. If the initial cost of equipment was high, operating costs were low. On the basis of outlay per square yard, hydraulic mining was cheaper than any other method.

A big disadvantage was that the technology left in its wake destruction far greater than that previously experienced by placer miners in the western United States. In California, because of the impact of river-borne debris, which was engulfing farmlands downstream, legal restrictions on hydraulicking in 1884 all but eliminated it as a serious mode of gold mining in that state.[8] Not so in Idaho: by the 1863 season in the Boise Basin, one group of miners had already installed a hydraulic giant on Elk Creek.[9] In early Idaho, all of these traditional placer methods and devices were used—the pan, the rocker, the long tom, sluicing, and hydraulic mining—often, when available, using mercury for amalgamation of the gold.

White miners were not alone in Idaho. The Chinese were very much in evidence. With hard times in their native land, countless numbers had been migrating to mining regions elsewhere in the world. In the West, they had appeared during the California Gold Rush and been brought in during the 1860s to help build the Central Pacific rail line. When the rich Idaho placer deposits were discovered, thousands of Chinese eventually came north to the new diggings, although white miners made it clear that the interlopers were not welcome. In a few years, once the easy ground was no longer producing much, the whites were happy to sell it to the newcomers. According to the 1870 US Census, there were 3,853 Chinese miners in the state and only 2,719 white miners.[10]

The Chinese were frugal, productive, and resourceful men who squeezed the most from their worked-out claims. Moreover, they brought old technology to the scene and adapted it to new uses. An example was the so-called chain-pallet pump, a device previously used to drain and irrigate Asian rice fields. Now it proved invaluable as a means of de-watering flooded ground or diverting water to a sluice on a higher level. Soon, the Chinese pump and new types of water wheels caught on among white miners as well.[11]

In many Idaho placer areas, the days of easy mining were short. Reachable deposits were soon exhausted; bedrock was deeper underground and beyond the current digging range. One alternative was to contrive a new technology to handle the deeper gravels efficiently in large quantities, applying economies of scale. Such equipment was usually lumped under the name *dredge*, a generic term that might mean any one of a number of mechanical systems that involved digging auriferous gravel, separating the gold from it, and discarding the waste. It is with these "dredges" and their application in the state of Idaho that this study is concerned.

Among the many innovations in mineral technology in the late nineteenth and early twentieth centuries, several would truly revolutionize the industry. Cyanidation and flotation would contribute quiet but deep changes in milling. The low-grade copper technique initiated by Daniel Jackling and others would bring bulk handling to previously useless ores, at a time when the new automotive and electrical industries were increasing their demand for that metal. Gold dredging would apply Henry Ford's mass production to placer deposits, again enabling the lucrative working of ground that was earlier untouchable.

In terms of numbers and the production of precious metal, the so-called bucket-line dredge was most important in the American West, including Idaho. It was depicted by an early contemporary as consisting of "a floating hull with a superstructure, a digging ladder, endless chain of digging buckets, screening apparatus, gold-saving devices, pumps and stacker. It could be described as a floating mill with the addition of an apparatus for excavating and elevating the ore."[12]

The first successful gold dredge in the United States was built in Montana. Constructed by the Bucyrus Steam Shovel and Dredge Company under the supervision of Samuel S. Harper, it was a bucket-line dredge for the Gold Dredge Company of Chicago. The boat slipped into the waters of Grasshopper Creek at Bannack in May 1895, after the wife of the firm's president, Herman Reiling, broke a bottle of champagne and christened it the *Fielding L. Graves*. This was the first workable machine of its kind in America. Bucyrus also built a second Montana dredge, the *A. F. Graeter*, the following year. Both Harper and Reiling subsequently played roles in Idaho's dredge mining history.[13]

Since the 1850s, Californians had tinkered with the idea of harnessing the long-established harbor dredge for digging gold, but despite a plethora of efforts—some bizarre—forty years later they had still not achieved their goal.[14] They had failed to build on each other's work. It was ingenious New Zealanders who led the dredge parade. At first they used current-wheel bucket-line machines run by paddle wheels driven by a swift current; next, by 1881, steam-powered dredges were doing the work. In 1895, the year of the first workable American dredge, New Zealand had eighty-four dredges in operation.[15] By that time, skilled technicians as well as the latest versions of dredges were being exported to the far corners of the mining world.

One of the technicians, Robert Postlethwaite, manager of the New Zealand Engineering Company, migrated in 1896 to San Francisco, where he became chief dredge designer for Risdon Iron Works. Risdon was one of the most important firms in developing mining machinery in the late nineteenth and early twentieth centuries. Postlethwaite's first Risdon bucket dredge was completed in 1897 and set to work on the Yuba River, but unfortunately it sank in the stream's turbulent water. The following year a second bucket dredge fared much better: built for Thomas Couch, a Montana miner of some means, that machine performed very well on the Feather River. It not only started Yuba Manufacturing Company's Wendell Hammon, Couch's partner, on his way to becoming a major figure in the gold mining industry; it also launched a large-scale dredging boom that totally revolutionized placer mining.[16]

In time, after much experimentation, the New Zealand and Montana approaches would come together to produce a hybrid California type vastly superior to either of the two earlier approaches. The awkward double-lift system of the Bannack boat gave way to Postlethwaite's single-lift arrangement, in which the bucket line carried the gravel to its highest point, eliminating the need for pumping the fine material an additional distance for screening or washing. Close-connected buckets of special steel were substituted for the open-connected type championed by Risdon, so that, linked one to another, they gathered the material loosened and dropped by previous buckets. The four-sided upper tumblers, which served as a huge sprocket to power the bucket line, now became pentagonal, hexagonal, or even round and were eventually cast in one piece, body and shaft together, weighing as much as twenty tons.

The smaller lower tumblers that kept the line in place at the bottom of the digging ladder also evolved, with a new type of steel alloy flanges and bearings for underwater use and high abrasion. First high-carbon and then chrome, vanadium, and especially manganese steel was used to lengthen the life of parts subject to wear. Time brought improved scrubbing and screening devices, huge revolving perforated cylinders called trommels (also called grizzlies), washed by jets of water. The double banks of gold-saving tables over which the fines were sluiced, which had replaced the New Zealand shaking screens, remained the most primitive part of the apparatus.[17]

Because of the need to dig tough terrain that might contain boulders, the California-type dredge had to be heavier and stronger. These machines abandoned the older system of cable headlines to keep their position and relied on steel or wood pointed spuds, massive posts dropped into the soil under the pond to provide a pivot around which the dredge could swing "like a dancer turning on her toe." Using one spud as an axis and manipulating the pressure on shorelines, the dredge chewed up the ground in an arc, seesawing along. To discharge the leavings—the tailings—the belt stacker became standard, replacing the early Montana machines that had used flumes and the Risdons, which had used a bucket conveyor to handle the debris. At first, the wooden hulls of the dredges were beautifully crafted by shipbuilders, then they were built by ordinary carpenters, but as more and more of the machines were exported to the tropics and the frozen North, steel hulls were used.[18] Because of its low cost and the versatility of variable-speed motors, electricity was the preferred source of power for dredges, either from individual generating plants or from local utility companies. Isolation often forced the use of steam or diesel engines, however.

By around 1910, it was clear that placer mining was about to be revolutionized and that the California-type bucket dredge would increasingly play a key role. Wendell Hammon's Yuba Manufacturing Company led the way, but the other major manufacturers fell in line: Bucyrus of south Milwaukee, the Marion (Ohio) Steam Shovel Company, the New York Engineering Company, and Risdon Iron Works, which had been absorbed by Union Iron Works and later by Bethlehem Shipbuilders. It was obvious that the boom in dredge construction was not merely domestic but global in scope. Wherever it was built in America and whether it was used at home or abroad, the California-type dredge came to be accepted as the standard of excellence.

Like others, Idaho dredge operators quickly learned to abandon the rivers themselves and to float their dredges in pits "inland," bringing in water using ditches or pumps to create a pond that moved forward with the dredge. The also quickly discovered the value of using the Keystone drill to dig test holes before sinking substantial capital into machinery or property. Both were admonitions of Robert Postlethwaite.[19] That caution does not mean there could not be failures: in the early years there were many as a result of poor judgment, incompetence, bad machinery, or chicanery. With careful testing, continuous operation, and a little luck, many bucket-dredge owners were able to make a profit from ground that would yield only ten or fifteen cents a cubic yard to individual miners. For repairs, any dredge enterprise of any size kept its own machine shop, with a stock of spare parts. Dredge operators hoped to work twenty-four hours a day but usually averaged about eighteen hours. Their schedule called for digging approximately 360 days of every year. Thus, dredgemasters kept a wary eye on the clock and the calendar and made a meticulous record of downtime, which might result from removal of a rock or a piece of timber or from bad weather tying up work for several days or, at worst, a month or two. In addition, if a dredge sank, not an unusual occurrence, it could be inoperable for weeks or even months.

As indicated earlier, the bucket dredge was not the only machine that bore the name *dredge* in the nomenclature of mining men and mining journals. (There was even a dry-land dredge.) Hydraulic dredges, better known as suction dredges, were used in various places throughout Idaho. Mounted on a flat scow, this equipment utilized a powerful centrifugal pump to suck material from the bottom of a stream, after which the gold was separated either on the scow or on a separate dredge. Modern, more sophisticated hydraulic dredges could also have a digging ladder attached to the suction pipe, with a motor-driven cutter head to chop and loosen the muck.[20] Such dredges had little success in Idaho; only one company that used a basic suction dredge ever paid a dividend in the state.

Any mechanized crane with a two-part bucket was a clamshell dredge; one with four parts was an orange peel dredge. Both were tried in the Idaho placer but to no avail—they lost too much gold. Early steam shovels, also called dredges, were extremely successful in the open-pit iron, copper, and coal mines but much less so in gold mining. When modernized with electricity or diesel fuel, they found a niche in some parts of the state for digging gravel and filling trucks, which they took to a separate washing plant—not always the most economical procedure.

Finally, in the Great Depression of the 1930s, after the federal government had raised the price of gold to thirty-five dollars an ounce, a worthy competitor appeared for the bucket-line machine. It came in the form of the dragline

FIGURE 0.1. Like clamshell and orange peel dredges, this ineffective dipper dredge failed to save gold, although it seems to have caught on with the smart set. Courtesy, Idaho State Historical Society, Boise, ISHS 72-112.b.

dredge, first devised by an imaginative Californian. Draglines were not new. They were well-known for moving dirt on construction sites, and during hard times many of them were standing idle. Earlier, they had been tried in placer mining but had failed. But now, when modernized especially for mining, they soon proved successful.

A dragline dredge consisted of two units—one for digging, the other for washing and processing the gravel. Digging was done by a self-propelled power shovel equipped with a scraper bucket suspended from a structural steel boom fifty or more feet in length. Buckets ranged in capacity from three-quarters of a cubic yard to three yards and were pulled toward the shovel, to be swung in any direction by a separate hoist drum on the machine. The second unit was one of three kinds of washing and gold-saving plants: stationary, movable, or floating. The stationary plant usually required trucking the gravel, an extra expense. The movable plant was on its own caterpillar

FIGURE 0.2. This dragline dredge was built in 1936 on Red Horse Creek, a tributary of the Red River. Workers are still constructing the washing plant. Courtesy, Idaho Gold Fields Historical Society, Smith Collection, Elk City, ID.

treads but was awkward. The most efficient was the floating plant. In general, all of the plants copied the machinery of the bucket dredge for separating the gold, but on a smaller scale.[21]

The dragline had several advantages. Initial cost was lower than that of the bucket dredge, and it was more maneuverable in working tight pieces of ground. Areas best suited for the dragline were shallow deposits too small for a bucket dredge and too low-grade to be worked by hand. But the dragline could not dig under water, and the bucket-line dredge was superior in saving gold, not to mention its much lower cost of handling a cubic yard of gravel.[22]

Idaho's first dredges, both bucket and hydraulic, came on the Snake River in a frantic boom from the 1880s to about 1910. Dozens of dredges were involved, and millions of dollars were spent in a futile attempt to save the river's fine gold. Successful gold dredging in the state started early in the twentieth century, in the high placers around Pierce, Florence, Elk City, Warren, Salmon City, Boise Basin, Stanley Basin, and Yankee Fork. The richness of these deposits had been skimmed by hand, then by various other processes, including hydraulicking. Most of these deposits experienced dredging efforts in the late 1890s, but not until the California-type machine became a reality did the bucket dredge really find its place in the Idaho mineral industry. When that happened, the state generally ranked fourth in the United States in the production of dredge gold, behind California, Alaska, and Montana.

Gold dredging was a messy business. The dredge, even at its best, was an ugly, graceless, tireless metal monster that clanked noisily and relentlessly along in its own dirty pool, tearing paydirt from bedrock twenty or thirty feet below the surface. At the same time, these seemingly insatiable leviathans left in their wake ruin and destruction, especially along the banks of streams where they left their tailings in huge windrows. John Gunther described a dredge that left behind "the kind of furrow that an enormous obscene un-house-broken worm might leave."[23] The damage to waterways inevitably brought efforts to regulate dredging in the late 1930s, with the farm element initially taking the lead and other groups, such as fish and wildlife interests, conservationists, and the press, gradually following suit. The Idaho Mining Association was a formidable foe, and it took until 1954 for a weak law to be enacted and another fifteen years before a statute that had teeth became law. By that time, gold dredging had virtually ended in the state.

Then came the aftermath: the cleanup. The aim was to return streams as much as possible to their native condition prior to dredging. Often, stream flow had been modified to eliminate meandering, leaving little cover for fish. The process of dredging also eradicated most of the vegetation on the banks, leaving no shade for salmon or other species. All types of dredges left tailings, sometimes in the waters, sometimes on the edges of streams. At first, restoration was haphazard; then, in 1980, a congressional law mandated that the Bonneville Power Administration spend a portion of its profits to mitigate the adverse effects of the Columbia River dam system on fish and wildlife in Idaho, Montana, Oregon, and Washington. This arrangement brought billions of dollars to renovate stream in Idaho alone, much to the benefit of several Native American tribes, although, as of 2014, work is still ongoing.

Notes

1. Harold Kirkemo, "Prospecting for Gold in the United States," US Geological Survey, January 10, 2013, http://pubs.usgs.gov/gip/prospect2/prospectgip.html.

2. Otis E. Young Jr., *Western Mining* (Norman: University of Oklahoma Press, 1970), 110; William Greever, *The Bonanza West* (Norman: University of Oklahoma Press, 1963), 47.

3. Greever, *Bonanza West*, 47.

4. Young, *Western Mining*, 113.

5. Greever, *Bonanza West*, 48; Thomas Henry Watkins, *Gold and Silver in the West* (Palo Alto, CA: American West, 1971), 176, 185.

6. Duane A. Smith, *Mining America* (Lawrence: University of Kansas Press, 1987), 26.

7. Robert L. Kelley, *Gold versus Grain* (Glendale, CA: Arthur H. Clark, 1959), 41.

8. See Rodman W. Paul, *California Gold* (Lincoln: University of Nebraska Press, 1947), 152–70.

9. Merle W. Wells, *Gold Camps and Silver Cities*, Bulletin 22, 2nd ed. (Moscow: Idaho Department of Lands, Bureau of Mines and Geology, 1983), 5.

10. Merle H. Wells, "Idaho's Chinese Pioneers," Idaho Historical Society Education Series no. 4, Boise, n.p.

11. Liping Zhu, "No Need to Rush: The Chinese, Placer Mining and the Western Environment," *Montana: The Magazine of Western History* 49 (Autumn 1999): 51.

12. Arthur Lakes Jr., "Gold Dredging Practice in the Placers of Colorado," *Mining Science* 59 (January 14, 1909): 28.

13. *Dillon (MT) Tribune*, May 17 and 31, 1895; *Dillon (MT) Examiner*, May 22, 1895; Bucyrus Company, "List of Dredges," 100075A–100075D, courtesy, David M. Lang, Bucyrus International, Inc.

14. See Clark C. Spence, "From Gold Pans to California Dredges," in *Riches for All*, edited by Kenneth N. Owens (Lincoln: University of Nebraska Press, 2002), 296–316.

15. Thomas A. Rickard, "Alluvial Mining in Otago," *Transactions of the American Institute of Mining Engineers* 21 (1892–93): 464; "The Mining Industry," *Appendix to the Journals of the House of Representatives of New Zealand* 1, no. C-2 (1889): 145; Robert McIntosh, *Three Prize Essays on the Present Condition and Future Prospects of the Mineral Resources of New Zealand, and the Best Means of Fostering Their Development* (Wellington, NZ: John McKay, Government Printer, 1907), 92.

16. John Barry, "Bucketing up the Wash: R. H. Postlethwaite and the Evolution of the California Style Gold Dredge" (paper delivered at the Buckets of Gold Conference, April 2006, Alexandria, New Zealand); *Who's Who in Engineering, 1922–23* (Brooklyn, NY: Lewis Historical Company, 1922), 1009; Robert H. Postlethwaite, "Menakher," *Mining and Scientific Press* 109 (September 19, 1914): 432; "Dredging for Gold in the Western States," *Overland Monthly* 42 (September 1903): 265.

17. Charles M. Romanowitz, "California's Gold Dredges," Mineral Information Service, California Division of Mines and Geology 23 (August 1970): 165; Norman Cleveland, "Bucket Line Dredge Methods," *Mining Congress Journal* 27 (November 1941): 20; Lewis E. Aubury, "Gold Dredging in California," *California State Mining Bureau Bulletin* 57 (Sacramento: California State Mining Bureau, 1910), 38, 40–41; Frank W. Griffin, "Reminiscences," *Western Mining News* (San Francisco) 3 (June 1935): 14; Frank W. Griffin, *Mining World* 24 (May 19, 1906): 607–8; *Mining and Metallurgy* 25 (May 1903): 461.

18. H. H. Dunn, "Eight-Cent Digger Earns Millions," *Popular Mechanics* 49 (January 1928): 77; "Increasing the Life of a Dredge," *Engineering and Mining Journal* 90 (July 30, 1910): 202; *Engineering and Mining Journal* 91 (March 4, 1911): 483; Aubury, "Gold Dredging in California," 54–57; George Walsh, "Economy in Mining with the Modern Gold Dredge," *Mining World* 24 (May 19, 1906): 608.

19. Jack R. Wagner, *Gold Mines of California: An Illustrated History of the Most Productive Mines* (Berkeley, CA: Howell-North Books, 1970), 41, 43; *Mining and Scientific Press* 78 (March 4, 1899): 227.

20. John H. Wells, "Placer Examination Principles and Practice," *Technical Bulletin* 4 (US Bureau of Land Management, Phoenix Training Center): 133.

21. E. D. Gardner and C. H. Johnson, "Dredging and Other Forms of Mechanical Handling of Gravel and Drift Mining," *Information Circular* 6788 (Washington, DC: US Bureau of Mines, 1935), 51.

22. Stuart S. Holland, compiler, "Dragline Dredging Methods," *British Columbia Department of Mines Bulletin* 16 (Victoria: Banfield, 1942): 1, 5, 29.

23. John Gunther, *Inside U.S.A.* (New York: Harper and Bros., 1947), 159.

I

Early Snake River Boom

From its origins on the Continental Divide in the Teton Range in western Wyoming, the Snake River flows southward in a large arc across southern Idaho, then forms that state's western border as far north as Lewiston before it meanders westward to its eventual confluence with the Columbia in southern Washington. Since as early as the 1850s, when soldiers at the old Fort Boise made limited discoveries, gold has been found along most of the 800-mile course through the Gem State. Optimistic bands of prospectors moved south to expand operations in the 1960s, without much success, but in the 1970s many of them located workable placer ground in scattered sites, among them the Shoshone Falls region, the area near J. Matt Taylor's bridge (which became Idaho Falls), the Hagerman Valley, below the Raft River, and west of American Falls. But these were scattered small-scale, labor-intensive operations that spawned no serious influx of capital or miners as long as a major obstacle existed. It was the nature of Snake River gold itself that provided the real challenge to profitable mining.[1]

DOI: 10.5876/9781607324751.c001

By 1880 Snake River gold had been recognized as constituting extremely fine particles—some so small that it took 1,000 to make up the value of a penny. Others were even more minute, and the 3,000 or 4,000 particles required to make up a penny's worth of gold could not be recognized at all: *invisible gold* was the miners' description and *micron gold* was a more modern term.[2] Early scientists believed the fine gold came from the waters of a Miocene lake, but modern geologists have determined that it came from Rocky Mountain deposits and that it originated as minute particles.[3] Scattered the length of the stream, this "flour" or "float" gold appeared in river bar deposits, as well as in "skim bars" and "bench gravels" on the bank. A skim bar was an ancient river bar, in which the gold was found near its top, while bench gravels were higher up and gold was found in streaks anywhere between bedrock and the surface.[4]

To many observers, the Snake River placers seemed a promising field for the gold dredge, a relatively new technology that had proved successful in New Zealand and which, after years of experimentation, was beginning to catch on in California, Montana, and other western states. The stream's current ran at a moderate rate of three miles an hour, severe flooding was rare, and the climate permitted operation for nine or ten months of the year. Best of all, according to optimistic reports, there were "oceans of gravel," which showed values of from thirty cents up to five dollars with a pan or a rocker. The belief that there was a wealth of gold in and along a great portion of the river's length was common, although few of the numerous miners since the 1860s were taking out more than day wages.[5] In 1899 Don Maguire, a self-taught Utah mining man, estimated that the gravels contained at least $2 billion in gold.[6] A popular Oregon writer noted a year later that "men, to-day, dredging the sandy bed and banks of the river Snake say there is fine gold enough in the drifts and bars alone to pay the national debt over and over again"—words Boise businessmen echoed enthusiastically.[7] Since pay-dirt was usually on or close to the top of the bars, deep dredging seemed unnecessary. At least that was what mining analysts of the late nineteenth and early twentieth centuries believed, thus setting the stage for a brief era of gold-dredging activity along the Snake. No wonder there were so many repeated efforts to apply a more advanced technology. Unfortunately, on the Snake it was a high-interest, low-success industry.

The period prior to about 1910 was largely a time of trial-and-error experimentation, often by self-styled experts testing their own inventive machines and processes. Some mining men welcomed attempts to apply modern technology; others dragged their feet, holding fast to the old ways of doing things. The new dredging ideas on the Snake extended from around Huntington in Oregon to east of Idaho Falls and were especially prominent in the region from Weiser to perhaps Caldwell, with lesser applications in the stretch from Grand View, Bruneau, and Glenns Ferry. Momentum on the river picked up again from Minidoka through American Falls, Blackfoot, and Idaho Falls. Only a single company organized for dredging on the Snake in the Jackson Hole region of Wyoming has been noted.[8]

One of the first efforts to utilize yet unproven dredge equipment on the river was first announced to the public at the end of 1881, with the arrival in Boise of four Detroit businessmen en route to Starrh's Ferry on the Snake. This quartet—George C. Fletcher, T. H. Champion, E. M. Hough, and John Shiri—had obtained ground near the mouth of Goose Creek and intended to apply the new technology there.[9] According to the *Salt Lake Tribune* that December, John F. Sanders of Ogden had investigated the Snake River placers beginning in 1870 and experimented with saving the fine gold and processing the black sands previously cast aside. He was convinced that machinery could be applied, and subsequently he patented the Sanders Ore Concentrator and the Sanders Patent Vacuum Dredge.[10] In addition, he held Letters of Patent on "Composition for Dissolving the Coating of Gold Ore," in essence a simple chemical formula for breaking down the black sands. The secret was sixteen parts of potassium cyanide to one part of glacial phosphoric acid, with the solution mixed in an iron barrel (or one made of some "other proper material") with enough water to form a pulp with the sand and gravel, then agitated in the container for fifteen minutes to an hour. The result would be gold pure enough to amalgamate with mercury.[11] (Was this a forerunner of the cyanide process?) Four companies were formed to operate between American Falls and Salmon Falls: one by investors from Detroit, another by Boston capitalists, a third with money from Detroit and Buffalo combined, and a fourth by men from Salt Lake City. It was contemplated that all machinery would be afloat and that some rigs would handle 100 tons of gravel in a ten-hour workday and others would handle twice that amount.[12]

Finally, in the spring of 1882, according to a correspondent for the Chicago *Mining Review*, two dredges built under Sanders's patents conducted initial trials near the Goose Creek ferry—a mile west of the modern town of Burley.

They were owned by two Detroit companies, the presidents of which were there with high hopes. Whether any of the four earlier visitors were included was not mentioned. The inventors had sold the scheme on the basis that the sands of the river contained from $500 to $2,000 worth of gold per ton. The machinery consisted of a vacuum drum for sucking the sand from the river bottom, with "innumerable shakers and washers," all set in motion by steam power. The result was clear-cut: either the dredges were a failure and saved no gold, or there was little gold to save. In this case, the sands averaged about 20 cents to the ton. The visiting corporate executives were highly indignant, "and their denunciation of the patentees were [*sic*] both ludicrous and amusing," according to one editor.[13]

Twenty years later, J. L. Mallery, a seasoned mining man employed by one of the four companies, contended that "the enterprise was a fraud from its inception." The vacuum dredge was antiquated, with equipment "entirely inadequate for saving gold in any form." Moreover, the business was based on fraudulent astronomical assays, some running as high as $20,000 a ton, according to Mallery, writing after the fact. The promoter claimed that iron grains of black sand enclosed a golden interior—"minute geodes of gold, we might say." This "invisible but fascinating proposition" was backed not only by the high "tests" but by the lure of platinum, tungsten, and tin, which "also played 'peek-a-boo' among the grains of magnetite."[14] That unfortunate venture cost the companies $80,000. Mallery does not name the promoter. Whether it was the inventor, John F. Sanders, we may never know, but whoever he was, when he left the scene he headed in the general direction of Ogden: "He, the promoter, stayed by his guns until the bristling array of failures made retreat imperative. So he decamped in the night, and was [spotted the] next morning by the stage driver [with] blankets on his back, steering south."[15]

Sometime later, in 1886, C. C. Hill, a Chicago mechanical engineer, described a special dredge he had designed some years before to handle the alluvial fines at an unidentified place along the Snake, where, he noted, "it has been the dream of a great many, both smart and stupid, men to get that gold."[16] Hill had been hired by two separate companies, made up of the same capitalists, that had rights to a bank twenty-six feet high from water level and extending four miles along the river. He devised a dredging machine with eight chains of buckets, which together would cut a swathe in the same fashion "as a man would start in a hay field to mow."[17] Each iron bucket was ten inches in diameter and shaped "very much like a tea-cup, and each was armed with a steel spike, the better to loosen the packed soil." Powered by a twelve horsepower steam engine and designed to move parallel along the river, chewing up the dirt, removing the gold, and dumping the debris into the stream, the machine had a capacity of about 1,500 tons of gravel in a ten-hour day. Hill believed the concept of the dredge was a success: "I could dredge the material well enough; I could wash it well enough; but after I had washed it I found that I could scarcely get enough to pay for the trouble."[18] Thus ended that endeavor, as would be the case with most subsequent efforts in the same waters. At the same time, miners using time-honored rockers and sluices were at least taking out wages.

In June 1885, when three men were lost when their rowboat capsized and washed over Salmon Falls, south of the present town of Hagerman, that tragedy was witnessed by twelve men "working on a ferry-boat for a New York placer mining company," according to a contemporary newspaper.[19] Perhaps it would not be a great stretch of the imagination to view this as a crude attempt at dredging.

A few years later, in 1890, the large "steam mining boat" *Gold Seeker*, owned by the Williams brothers of Payette, operated along the banks of the Snake near Nyssa. Positive reports indicated that it was doing well and managing to save the fine gold. Soon, however, its owners clashed with the Union Pacific's Shortline Railroad. The *Gold Seeker* had worked out a short strip between two of the Oregon Shortline's bridges but was unable to continue because of them. The brothers first petitioned the railroad company to build a "draw" section that could be raised to allow passage of the dredge. When the Union Pacific refused, arguing that the cost would be $200,000 and that the Snake was not a navigable waterway, the Williams brothers filed a lawsuit, apparently to no avail. This setback seems to have brought the enterprise to halt.[20]

Nonetheless, Idaho newspapers continued to laud the richness of these sands and predicted a rosy future, once a cheap method of handling them and saving the precious metal had been perfected.[21] Numerous early suction dredges and centrifugal pump machines were tried, most of which were as unrewarding as the Detroit fiasco; and a good deal of capital went into a variety of patented amalgamators, concentrators, grizzlies, and shaking tables, few of which proved very effective.

Snake River gold had received a great deal of publicity prior to the early 1890s, its production was on a small scale and relatively limited. Excitement was easily misplaced. Those in touch with the reality of the region characterized the gold particles as "so lightweight as to be illusive." As a writer in Caldwell noted, a man with a rocker might make $1.50 a day in the summer, but except for a few scattered bars between Shoshone Falls and the Oregon border, the flour gold was not encouraging. Numerous sites along the way might have been profitable if water could have been applied for sluicing or hydraulicking, but the gradual fall of the stream and the level outlying terrain made this difficult. Large-scale work with experimental equipment seemed to many to be the answer, but hopes were generally dashed as quickly as they rose. "The great excitement caused by the trial of a new machine, said to catch the invisible gold, has subsided," wrote one Idahoan in early 1894 of an unidentified and unsuccessful contraption, then he went on to deny its need. "New-fangled gold savers are of secondary consideration," he said, "for there are dozens of home-made machines that will save 90 percent of the gold, and the cost to build is a little lumber, nails and labor."[22]

This may have been a snide oblique reference to Thornton Williams's "big floating gold-saving dredge," recently built on the Snake a few miles south of Payette. The sixty-five-foot by twenty-foot self-navigating stern-wheel flatboat was powered by a thirty horsepower marine engine and boiler, and its endless chain of forty-eight buckets, capable of holding about twenty pounds each of sand and gravel, could dig either on the river bottom or in bars on the bank. Gold was separated by a special agitator and steam rocker and recovered with quicksilver from copper plates, while tailings went off into sluices. "The gravel is worked so thoroughly that no gold escapes in the river," the owner reported. Nine months after it was completed, this example of "Steamboating for Gold," as the press called it, was still in operation—at least intermittently—but it is dubious that its three-man crew reached the owner's expectations of taking out $100 a day, and its working life seemed to have been near its end.[23]

In May 1894 Chicago capitalists Root Brothers and Dunbar announced plans to install a suction dredge on the Snake between Shoshone Falls and Blue Lake. A little later that year Tom Ball, a Pocatello railroad conductor, was trying to finance a dredge with a capacity of 1,000 tons a day, to integrate with a gold-saving machine he had invented. Early the next year, the Blue Mining

Company did put in some kind of dredge twenty-five miles from Shoshone.[24] If or how long it may have operated is unclear. Little is also known about the machinery installed at about the same time on Mann's placers near American Falls by the Roth Amalgamating Company. This consisted of a steam engine, boiler, pumps, grizzlies, and two Roth amalgamators on a flatboat, working to dredge up gold-laden black sands from the river bottom.[25] A little later, in August 1895, the Red Star Mining Company built a sixty-foot by eighteen-foot suction dredge, "embodying the best features of five different mining processes," a few miles below Payette.[26] Further mention of that enterprise does not appear, and it, too, likely "went up the spout," as nineteenth-century British would say of a failure. Nor do we know whether E. H. Lopez of Boise, a man with some dredging experience, succeeded in his plans to build a steam-powered dredge on the Snake near the Blackfoot bridge.[27]

The exception to the many rigs with disappointing results was the Sweetser-Burroughs dredge, which in 1894 went into the Snake in south-central Idaho, near the later location of the town of Burley. Lewis Sweetser and the Burroughs brothers, George and Harry, were "local boys," small-time ranchers in the area who also happened to be engineers. Their machine suction dredge was remarkably successful and worked for a number of years. It accomplished something no other boat on the river managed to do—pay a dividend. The men were so pleased that they bought a five-foot bucket-line dredge, which proved to be much less successful.[28]

One of the most innovative dredge operators on the early Snake was William W. Priestly, a pioneer irrigator in the Hagerman Valley. In 1894 Priestly had installed a brilliant pneumatic ram system using the falling water from Thousand Springs to create air pressure to lift river water up seventy-five feet to a ditch for use in the fields. No motors of any kind were used, and similar compressors were subsequently employed elsewhere in Canada and the United States at the turn of the twentieth century.[29]

In 1896 in the same area, Priestly built a gold dredge, to which he applied the same principle. In late June one bystander described the twenty-four-foot by sixty-five-foot boat as the inventor started it up, calling it a "wonderful piece of machinery":

> Soon the water came thundering down the large pipe into the tank, or air separator where it escapes out a short pipe 40 feet high, and [is] thrown 150 feet into the air, falling on the sides, making a complete and beautiful geyser.

> Soon the rumbling noise in the air pipe gave the signal that the air was fast approaching the boat. This immense dredge has 16 mining tables 20 feet long by 3 feet wide, in another instant they were covered with water and gravel from the river . . . If you Shoshone people come in contact with a "sly coon" or a cloud burst you may know Mr. Priestly has got mad and turned things loose.[30]

By August, Priestly had doubled the size of his crew, and a month later he was also using his pneumatic pump to run a hydraulic monitor. Like most of his contemporaries with more conventional suction or bucket-line dredges, he was stymied by the inability to save the fine gold of the Snake. In the end, his boat was auctioned off for $500 at a sheriff's sale. He promptly bought it back from the buyer and retired to his irrigated fields.[31]

Still, optimists continued to finance elaborate and often more absurd dredge machines. In early 1896 A. H. Butler completed a large barge for E. H. Lewis at Forman's Ferry, fifteen miles from Caldwell. Equipped with steam machinery "of all sizes and description" and including good boarding and lodging facilities for up to fifty workers, the dredge was described as "a veritable mining camp."[32] If it worked, there is no record, although more than two years later a large unidentified dredge was reported operating near Caldwell.[33] In the spring of 1896, two Emmett residents reported a good deal of activity near the Ontario ferry on the Snake and the fact that the Williams brothers of Payette were back. Two large steam-powered suction dredges were under construction, one of which was for the brothers. Machinery was already at the Ontario railroad station, ready for installation when the hulls were complete.[34]

In the summer of 1897, Butler put another boat on the Snake near Nyssa for Baker and Scott of Tacoma. This was the largest boat yet built on the river: it had two fourteen-inch centrifugal pumps and a plant that would handle 250 yards per hour.[35] At about the same time, a Lewiston newspaper reported that local entrepreneurs were investing in a new "Floating Gold Mill," a combination suction dredge, with powerful machinery that could throw a jet of water with such force "that it will tear down a mountain." This was no doubt the machine of the Hydraulic Placer Syndicate, a Nebraska concern with nominal capital of $50,000, built to work in the Snake at the John Weiss bar eight miles below Lewiston.[36]

Frederick Powell graduated from Columbia University School of Mines in 1883. After a variety of engineering jobs in Minnesota, Canada, North Carolina, and Georgia, he subsequently went to Idaho in 1897 as manager of the Standard Gold Recovery Company, operating a gold dredge on the Snake River. The precise location of the Standard machine is not clear, but early the following year he displayed sixty-five dollars in bullion as a result of a four-hour test run of his "dredge pumping plant" below Caldwell. Powell was so encouraged that he increased the capacity of his machine. At that time he also announced that the Payette Alluvial Gold Mining Company would soon commence work on the Snake with a large dredge near Ontario. The fact that dredging was catching on was further demonstrated when Isaac Hammond agreed to have a big boat running on the Tietsort ground above Nyssa by May 1 and the Singer company anticipated having its dredge operational on the Midas bar before the season ended.[37] More than a year later, now employed by the Hammond Manufacturing Company of Portland, Powell was handling the maiden run of the Gold Giant dredge at Bridge Island, near Nyssa, and believed he had shown that Snake River gravel, averaging eight cents a cubic yard, could be handled profitably.[38] He continued with Hammond for four more years, according to his account, "furnishing and erecting complete plants," especially gold dredges, before leaving the corporate world to continue as an independent consultant on mining machinery.[39]

In July 1897 D. S. Richards and A. A. Richards of Boise transferred sixty acres of placer ground near Nyssa to an unidentified Portland group, which had already begun construction of a dredge. The Richardses also bonded the rest of their holdings on the river—around 2,000 acres—to Arthur Grothe, a transplanted Englishman residing in Payette, who had earlier been manager of Poorman Mining Company, Ltd., a British firm operating in Silver City. After the Richards transaction, Grothe immediately headed to London to organize a limited liability company to work the property on a large scale. Five months later, newspapers announced that he had erected "a great dredge boat" on the lower river near Ontario that would soon be in operation. There was speculation as to whether Grothe, an "eminent mining engineer," might adapt the new, highly publicized gold-saving invention of Professor Amos T. Fox of Presbyterian College in Washington State, but there is no evidence that he did. Further, there is no indication of whether his capital was local or came from abroad; nor do we know if there were any positive results.[40]

Andrew J. Wiley, a young Boise engineer who would subsequently have a stellar career in hydrology and in the electrical power field, successfully operated a suction dredge in the Snake near Grand View for a brief time in the summer of 1898. At the same time, Wiley was superintendent and chief engineer of the Owyhee Land and Irrigation Company.[41] Wiley believed the operation's profitability demonstrated his theory that the best pay was to be found at considerable depth—"below the bottom of the pit, though not beyond our reach." The real wealth was below the riverbed, where the gold was much coarser than the surface deposits.[42] Wiley's efforts did not last long.

Idaho's adjutant-general, John L. Weaver, would be the driving force behind several attempts to develop viable dredge operations in the early twentieth century. Born in Indiana, Weaver was educated in colleges in that state and was a lawyer and an ordained minister. In 1895 he came to Boise as pastor of the Christian Church. Although long a staunch Republican, Weaver broke ranks with the party over the free silver issue and supported William Jennings Bryan in the 1896 campaign; he continued to support the Democratic Party thereafter. Two years later, Governor Frank Steunenberg appointed him to fill the vacancy created by the resignation of General B. W. Figgins, who joined the volunteer troops in the Philippines.[43]

In the summer of 1900, Weaver headed a group with the intention of opening large-scale placer properties in the Stanley Basin. The plan was to tap Valley Creek at the base of Buckskin Mountain and run a lengthy ditch to its ground. Weaver also contemplated installing a dredge the next season,[44] but that part of the venture seems not to have materialized. Two years later he had changed his focus to the Snake River, not far from Wiley's earlier endeavor. Weaver purchased a large dredge in Salt Lake City, a machine that "had undergone a thorough and successful test for saving flour gold" in southern Utah, then transported it in seven railroad cars to Mountain Home, from whence it was hauled by Helfrich Mercantile the remaining twenty miles to the Snake River, opposite the town of Grand View. There, it was to be assembled for a new trial run within sixty to ninety days. Weighing over 200 tons and electrically powered by a 200 horsepower steam engine and dynamo, the machine featured a special V-shaped tank with a unique powerful spray mechanism designed to save the elusive fine gold. The large tank was lined with more than 480 amalgamating plates and pockets, twelve inches to eighteen inches long, into which fully 500 pounds of quicksilver were placed. A circular motion akin to that of a miner panning gold allowed the precious metal to settle on the mercury, and jets of water under 60 pounds of pressure played up from the bottom repeatedly to wash out the debris while forcing the fine grains to stick on the amalgam plates. It was estimated that each particle of gold was forced into contact with the plates 1,000 times. If it worked as promised, Weaver believed the solution to the Snake's placer problem was at hand, "and there would be millions in it."[45]

Several early test runs of the dredge were encouraging, and when the general went to New York in search of more capital, he took amalgam with him that he said represented an average return of $8.65 a cubic yard, a profitable figure indeed. Meanwhile, his manager ran the machine whenever he could obtain enough coal, which was difficult since the rig was said to burn several tons every twenty-four hours. Some members of the press challenged his estimates, however, pointing to irregularities in calculating yardage from the pit and to the presence of yet-unknown metals in the amalgam. The fact was that the dredge had operated for only a short period and did not prove successful. Because of defective parts, it could handle only a small amount of gravel per day. Thus, in the summer of 1905, after a hiatus of nearly two years, Weaver was back with new funding and a new plan of attack.[46]

The fact that Weaver had not achieved his millions was apparent when he appeared again on the mining scene in 1906 and started up a new dredge in the river, this time near Bruneau. It was reported that the general had spent at least fifteen years tinkering with his machine, and at least two of his patented innovations were incorporated into this model.[47] It was a bucket-line rig that was supposed to be able to handle one cubic yard of gravel per minute. Equipped with a special rotary screen, it dropped coarse gold, if any, into ordinary riffled sluice boxes for recovery. As before, the fine gold was separated by washing it with Weaver's improved patented water jets in a large tank. But Weaver was no doubt reacting to the US Geological Survey report the previous year on the value of black sands for other metals as well as gold when he provided for the magnetic removal of iron from the black sands that constituted the resulting concentrates. The entire operation was powered by two electric motors, one 64 horsepower engine and the other 125 horsepower, with the electricity generated by a steam engine–driven dynamo. By this time, Weaver was president and manager of the Utility Power Company, and his aspirations included a massive power plant at nearby Crane Falls, which would not only operate multiple dredges but also pump water for irrigation, electric lights, and general usage. Newspapers of the time mention two dredges, both designed by the general, but describe only one. Both would have saved the finest flour gold, and eight more had been ordered. According to the press, it was a successful enterprise that would revolutionize placer technology: "It will bring untold capital here, give men employment and furnish electric power for any and all kinds of business." With those optimistic thoughts in mind, Weaver headed east again, hoping to raise more funds.[48]

Although several publications reported favorably on the test run of the new Utility Power Company dredge, said to have cost about $100,000, it was soon clear that there were difficulties. The machine's operational life was short, and, for whatever reason, the company was out of business by 1907. The firm sold and transferred all of its property to the Imperial Dredge and Exploration Company, a New York outfit officed in Mountain Home.[49] The Panic of 1907 hit hard, and heavy debt was no doubt one cause of the transaction, along with a fundamental flaw: Weaver's patented amalgamating tank failed to capture the elusive Snake River fines. After his failure, Benjamin A. Dare, a New York investor and former director of the Utility Power Company, attempted a reorganization, with plans to utilize a patent dredge with an orange peel bucket, an effort that came to naught.[50]

In actual practice, when the power plant was built at Crane Falls, it was primarily to pump irrigation water to new agricultural lands. The general was part of it, but his cutting-edge dredges were not. They appear not to have lived up to their promise, and Weaver's name was never inscribed on the roll of successful pioneers in the field. Rather, his impact was of a different sort. Through Utility Power in 1906, he had already created an electric plant on the banks of the Snake, with a sizable pumping station capable of irrigating 20,000 acres of land on the north side of the river. More than 3,000 acres had been filed on as desert claims and water rights purchased from General Weaver.[51]

There was much activity on the Snake at the Midas bar site at the turn of the twentieth century. Located about twenty-four miles above Ontario, on the Oregon side at Big Bend, the bar was nine miles long and nearly half a mile wide, with an average depth of gravel of twenty-seven feet above the normal water level. In the spring of 1898, E. S. McComas of Baker City, the owner of considerable land along the stream, met at the bar with Captain C. H. Thompson of Spokane and John A. Manly of Grand Forks, British

Columbia, for a trial run of a prospecting boat the two men owned. The results agreed with those of various experts who had previously tested the ground, and Manly was so impressed that he headed for San Francisco to buy a dredge from Risdon Iron Works, one of the pioneers in the field.

Indeed, Robert Postlethwaite, the head of Risdon's dredge department, personally visited the Boise Basin and the Snake River in hopes of putting two boats on the latter and one in the former. He succeeded in the basin and installed a machine on Warm Springs property near Idaho City. Two other operators above McComas's property were doing well with a pair of drag scrapers and sold two of their claims for $5,000 in cash to Lyon and Company, which planned to move its Colorado River dredge to Midas immediately. O'Brien and Company was already erecting a dredge on Morton's Island near Ontario, and a New Zealand outfit was building a big clamshell dredge near Nyssa. McComas had also leased ground to men who planned to construct three dredges with at least a 2,000-yard capacity per day each, and a few months later he was in Portland arranging for shipment of about 500,000 board feet of red fir lumber, to be used to construct seven dredges on the Snake. The Baker City newspaper reported that in the thirty miles between Payette and Great Bend, eleven mining boats previously built as "sandsuckers" were being converted to bucket-line machines.[52]

Soon, in Weiser, a US postal inspector was in town gathering information about the activities of a suspected con man, A. H. Wilcox of New York City, suspected of fleecing hundreds of naive investors who answered his "How to Get Rich" advertisements and bought worthless stock in the New Era and Gold Giant Hydraulic Dredging Companies, supposedly to operate at the Midas bar on the Snake.[53] McComas himself later reported that he had been "a pioneer in promoting and building" two dredges, the *Gold Giant* and the *Kankakee*, "only to draw my dividends in experience and resolutions 'never to do so any more,' " but he still believed mining the Snake could be a profitable venture, if only because of the value of the black sand.[54]

Activity in the Nyssa-Ontario stretch of the river continued to draw attention. In early 1900 R. G. Wilson of Salt Lake City made a trip to the Parma area, on the Idaho side, to view placer ground in which he was interested, with some eastern investors. He noted that two or three dredges had been operating successfully there but that greater dredging capacity was needed before the region could be compared in any way to the Boise Basin or the

Montana placers. Still, despite the fineness of the gold and its spottiness in the bars, Wilson believed adding more and larger boats would return very substantial profits to the companies then operating. He looked for a heavy influx of capital in the district the next season if some potential projects were brought to fruition.[55]

In the summer of 1899, another onlooker believed a solution had been found for retrieving the elusive fine gold—"so light that when dry it floats on top of the water like so much bran."[56] According to the press, a bucket dredge running day and night at Bridge Island, near Payette, was succeeding where all previous machines had failed, wasting hundreds of thousands of dollars and leaving countless wrecks along the river in the process. Cutting a channel twelve feet deep across the island itself, the new boat was flourishing and was able to raise black sand and fine gold to its sluices and separate the two by a new process, which was not described. The average value per cubic yard was 12½ cents, and the dredge was said to be clearing "the neat sum of $200 per day."[57]

Another new bucket dredge, owned by Baker City residents, began operations in the spring of 1900 about a dozen miles above the spot where the Boise River emptied into the Snake. The machine had an unusual apparatus for capturing the precious metal. Its sluice was shaped like a horseshoe around the inside of the boat, with special side mini-sluices to divert both gold and black sands to an old-fashioned Spanish *arastra* (perimeter mill for grading and diversifying gold and silver base) at the back of the hull. When thus suitably ground, the gold was amalgamated by the use of more quicksilver than was normally the case. The owners claimed to have saved 85 percent of the values.[58] How effective this machine really was is not known, but it caught the eye of the mining community.

When C. S. Crossman was interviewed by newsmen in Caldwell in the summer of 1900, his dredge—characterized as a "big gold-saving machine"—was at Bridge Island, about two miles west of Parma. Crossman had eight men at work and was running two shifts at a cost of $1.50 an hour, handling about 1,600 feet every twenty-four hours. He now believed he had solved the problem of catching the elusive fines of the Snake. Using what he called a gravity system that saved 80 percent of the values before it reached the burlap tables, he used a patented "liquid preparation" to capture the remaining 20 percent. In this process the gold, along with the black sands, passed over a burlap table with about two inches of water, then went to a second table where the sand and small gold particles that were coated with silica were treated to a mercury bath that kept them in constant motion until they amalgamated. The average value per cubic yard was said to be 12½ cents, and the dredge was clearing $200 per day.[59]

In 1900, when the future Idaho poet and essayist Paul E. Tracy was about eleven years old, his family moved from Silver City to the old Nat Graves "ranch," near the spot where Succor Creek (commonly called "Sucker" Creek) joined the Snake River—the present site of Homedale.[60] A couple of years later, his father led the family on a Sunday walking excursion to a dormant steam-powered dredge tied up on the river about two miles away. It was, in Tracy's words,

> A solitary red and useless dredge
> With its tall black stack
> Rising above the grey sage
> [It] floated for years and chafed a willow bank
> Along the Snake river.[61]

The company that owned the dredge had lured investors by emphasizing the richness of the river's bars—"enough gold in the gravels to pay off the national debt"—but its working life was brief. In Tracy's pungent prose, "It had chewed up a bar, spewed it out and called it a day." Now it was in the hands of a bushy-bearded, tobacco-chewing caretaker named Bradshaw, whose job was to keep the boat afloat, presumably for the better days that never came. He tightened loose cables, pumped out water inside the hull, and sealed cracks with oakum. As Bradshaw explained to the Tracy family, the entire business was a swindle "that never paid a red cent! The flour gold—if there was any, just went hell-bent-for-election over the riffles."[62]

A number of the dredges of this period were built by the Hammond Manufacturing Company of Portland, Oregon. Its proprietor, Isaac B. Hammond, produced not only bucket-line dredges for use on the Snake but others for the Alaska goldfields. Hammond was also aligned with promoter-investor and future Oregon senator Jonathan Bourne Jr. in efforts to attract capital to dredge operations in several areas. The *Yale,* the new Sweetser-Burroughs machine, differed from other Hammond-built rigs in both its screening apparatus and the arrangement of the burlap sluices. Originally,

Hammond's dredges had used Cariboo riffles for gold saving, but they failed to capture the fine Snake River gold particles. This problem was solved on the *Yale* by the substitution of a twenty-two-foot grizzly with small holes through which the gravel was screened and discharged by sluice to a shaking screen, which delivered the larger material to a stacker or tailings elevator. The sand fell under the screen evenly into four compartments in a large box, from which it was carried by water to a burlap sluice arranged in two banks on both sides. Engineer Robert Stanton visited a Hammond machine on the Snake near Lannon's Spur, not far from Parma, Idaho, in late 1899 and was impressed with both its construction and its operation. Its gold-saving mechanism was "almost perfect," according to the crew. Stanton said this Hammond dredge was owned by the Gold Giant Dredging Company and managed by L. C. Crossman "of Nyssa and Chicago." He indicated that its major original deficiency was the lack of a grizzly and that the company took the dredge off Hammond's hands as a compromise. After experimentation, the dredge "evolved [into] the best set of sluices & tables for fine gold that I have ever seen."[63] Some of the Hammond rigs were for use along the Idaho-Oregon border, but the Sweetser-Burroughs *Yale* worked the midsection of the state, and one operated by S. H. Abbott was still farther east, near American Falls—a town that also enjoyed a brief flurry of dredge building early in the twentieth century.

In mid-April 1901, when Abbott was in Salt Lake City buying a new shaft for his dredge on the Snake, he was said to be from Minidoka and had brought "some nice bars of gold" as tangible evidence of his successful operation.[64] Since Abbott was an experienced dredge owner who had run dredges at both American Falls and Minidoka, it is likely that one of those areas is what he had in mind. Earlier that same year, the Pioneer Dredging Company was also said to be operating on the Snake near Market Lake, one of the old names for Idaho Falls. With headquarters in Salt Lake City, the firm had been incorporated with nominal capital of $100,000. The main figures involved seem to have been J. D. Wood, F. J. Hagenbarth, and Jake Bamberger.[65] The corporate name does not appear again, so the company was probably short-lived.

In February 1902 the Acme Gold Dredging Company, headquartered in Pocatello, was incorporated. With nominal capital of $100,000 in shares of $1 each, $65,000 of its stock had already been subscribed. Its five directors were

S. H. Abbott of Minidoka (48,000 shares), C. K. McCoy of Salt Lake City (10,000 shares), Fred L. Presley of Salt Lake City (2,000 shares), Thomas F. Ferrell of Pocatello (3,333 shares), and J. J. Guheen of Pocatello (1,667 shares). The company's mission, as stated in its articles of incorporation, was "to carry on the business of gold mining on the Snake river, within the state of Idaho," but it cited no particular locale. A later publication, *Mineral Resources of the United States*, which summed up dredging activity in Idaho for many years following 1897, indicates that Acme had a boat somewhere on the river in 1902 but not in any other year.[66]

Also in February 1902, H. W. Muhleisen, who operated a dredge in American Falls, found his bucket-line dredge unsatisfactory and was in Salt Lake City conferring with machinery dealers, with an eye to substituting a centrifugal sand pump. Muhleisen reported that he was using a Pollock riffle, which worked well. He also noted that A. H. Tarbet and several others were building dredges on the Snake in the same vicinity.[67] Two months later the *Salt Lake Telegram* reported that a party of Chicago capitalists was headed for American Falls in a private railroad car to consider a dredge proposition there on the Snake.[68]

On both sides of the Snake near the mouth of the Blackfoot River, a good deal of gold had been taken from terrace gravel, but several efforts to utilize dredges and suction pumps had failed.[69] As early as April 1901, A. J. Schultz, superintendent of the Colorado Gold Mining Company, a Detroit outfit, reported that he had received shipment of a large dredge from Portland, no doubt manufactured by Isaac Hammond. The machine would have a capacity of 2,000 yards per day and would be put in place on placer grounds owned by J. G. Walsh, just west of Blackfoot. If it proved profitable, the company contemplated adding other dredges on the large amount of placer ground available.[70]

In May 1903 the incorporation of the Idaho Gold Dredging Company was publicized, with a modest capital stock of $25,000. Directors included Theron Geddes, who was also president, Warren Bogue, Schyler Shelp, Stephen O. Snyder, and Frank B. Stephens—all of Salt Lake City. Less than two months later that city's *Tribune* reported that the company had produced its first ingot of gold from the dredge on the Snake somewhere near Blackfoot. Under the supervision of Stephen Snyder, the start of the firm's big machine had been delayed by high water and a good deal of preparatory work, but first-run prospects were encouraging. "While the bar is not a weighty one," said the

Tribune, "it shows that the wealth of the waters is finally surrendering." But the boat did not appear on subsequent listings of active dredges, and no more was heard of this concern.[71]

By 1904 two suction dredges were digging in the area but were doing poorly—the result of a lack of knowledge and experience in both design and operation, according to the Idaho state mining inspector, Robert N. Bell, a vocal advocate of the suction dredge. Apparently the dredges' owners, the Blackfoot Construction Company and the Snake River Valley Gold Dredging Company, sought to consolidate and to bring in electric power from American Falls to replace steam. But these and other boats soon ran into unexpected trouble. Alfred R. Schultz, a geologist who visited the area in 1906, noted that in the same area "a well-equipped dredge," which usually ran on Walsh ground, had not operated during the previous season. Another company, a Portland firm with considerable machinery and a small dredge, was also not working. In addition, in the same area, competition for water between farmers and miners was growing.[72]

Construction of the state and privately funded Milner dam was completed in 1905, and the new Minidoka dam, the first Federal Reclamation project in Idaho, was finished the following year. A vast acreage was thus thrown open for irrigated farming. That coincided with an exceptionally dry summer, and especially upriver in the Blackfoot area, the riverbed went dry for a ten-mile stretch. Nonetheless, Schultz believed the old course of the stream, especially Diamond bar, showed excellent values in fine gold suitable for dredging and noted that one company had already purchased claims and planned to install one or the other of the machines. But while the experts who had examined the ground believed the conditions were ideal for dredging, success would hinge on "a large enough plant and intelligent handling . . . and improved methods of recovery."[73]

On the east side of Idaho, the early twentieth century also saw a number of dredging efforts in the Snake, some by experienced mining men, some by rank amateurs in the field. For example, Dr. Theodore R. Heintz, who ran an assay office in Salt Lake City, returned to his hometown just before Christmas in 1900 with "a nice retort of gold"—the result of a short run by the dredge of the Heintz Gold Extraction Company. Heintz was manager, secretary, and treasurer of the private corporation, while J. R. Hayes, proprietor of the Wayne Hotel in Detroit, served as its president. The company owned several hundred acres on the Snake River close to the town of La Belle, near Rugby, and had a boat capable of handling 1,000 tons of material a day. The gold was to be saved by an electrolytic process invented by Dr. Heintz himself, which he assured the press was "a complete success, as the recovery is absolute, not hardly a fraction being lost in the operation."[74]

The dredge had started up the previous month but had shut down briefly for minor changes. Since that time, it had been digging steadily in ground that ranged from thirty cents to a dollar and a half a ton, at an operating cost of five cents a ton, according to Heintz. It was expected that the boat would run all winter, with a labor force of twelve men in two shifts under the guidance of Superintendent J. H. Korty, who had owned an assay office and a laboratory on West Temple Street in Salt Lake City. Meanwhile, Heintz busied himself with plans and specifications for a second dredge, machinery to be supplied by local businesses and ready to go in the spring of 1901.[75] That was not to be. Within less than a month, the first boat burned to the water's edge and was a complete loss. Heintz vowed to rebuild at once and announced that it would take more than a fire to compel him and his colleagues to abandon the rich gold at La Belle.[76] He went to Detroit for a meeting of his corporate directors and visited New York and Chicago, as well as other eastern cities, presumably to raise capital. When he returned home he announced that the company would "begin a line of gold dredges to be put in operation along the Snake River." The appropriate contracts would be let after September 15, at which time the directors would assemble in Salt Lake City to inspect the workings of the Heintz Gold Extraction Company's ore-testing plant, which would be put in operation.[77] There is no indication that the good doctor was able to follow through with these hopes.

At about the same time, it is doubtful that another dredge venture in the same general area lasted much longer. Late in July 1900, President Rasmus Hoff and Manager William T. Chantland of the Upper Snake River Mining Company were visiting Salt Lake City from Idaho Falls, where the company owned a large tract of placer ground it was preparing to work. Part of their mission was likely to regroup and raise additional capital to equip their property with a dredge, which they hoped would be in place by the next fall. Thus, the Upper Snake organization was soon replaced by the Hoff Dredging and Exploring Company. Hoff was still president and the new corporation had a local flavor—most of the directors were from Idaho Falls—but much of the financial backing seems to have come from a Salt Lake City group led by Judge Clesson Kinney. The company proceeded to test about a third of its 1,000 acres and concluded that it averaged 17 cents in profit per yard and that with the $15,000 suction dredge on order, it would return a profit of $19,420 for a ten-month run. The boat made its debut in late October 1901 on the Snake at Basalt, about a dozen miles downstream from Idaho Falls, amid accolades from the press: the company would "undoubtedly have the best equipped dredging plant in the west," ground enough to last fifty years, and "sanguine hopes to [be] taking out 'dead oodles' of gold." About a month later, Hoff was still optimistic. A few alterations had been made, including one that would enable the rig to handle a good deal more gravel than anticipated. One problem had been that water that upped from the river into the big boilers was so cold that it was difficult to maintain steam. That issue had been solved by adding a water heater using exhaust steam from the engine.[78] Like most other large-scale ventures on the Snake, however, the dream was never realized. If the dredge dug at all, it was only briefly.

Who built the early dredges along the Snake? No doubt many were conceived and constructed by inventors in local foundries and machine shops scattered in major western centers—Boise, Denver, Salt Lake City—and a few came from as far away as Chicago and New York. Unlike dredges for Montana, Colorado, or the Boise Basin in the same period, they were not made by traditional producers of excavating machinery, like Risdon Iron Works in San Francisco and Bucyrus in south Milwaukee. Of the numerous boats on the Snake, there is no hard evidence that there was a single Risdon; records of Bucyrus for the years 1888 to 1950 indicate that none of the company's dredges went to a firm on the Snake. One did go to an outfit called the Snake River Mining Company in 1940, which worked on the Yankee Fork of the Salmon. By contrast, Risdon provided at least five new machines for the Boise Basin, while Bucyrus supplied two between 1898 and 1905. Colorado had one Risdon and five Bucyrus boats in the same period.[79]

The Hammond Manufacturing Company of Portland built at least four bucket dredges for the river, and its owner-president for a quarter of a century, Isaac Hammond, had a reputation for producing excellent machinery. Many of the dredges on the Snake were abandoned by their original owners after they failed to return profits, although a few were re-sold to new operators. The Hammond-built *Gold Giant* was once owned by E. S. McComas, then by the Gold Giant Dredging Company, before being acquired by the Snake River Mining Company. The Empire State Dredging Company bought the Sweetser-Burroughs workhorse suction dredge in 1906 and re-fitted it with new machinery in an unsuccessful effort to work the Diamond bar in Minidoka.[80] One entrepreneur, General Weaver, purchased a used dredge in Utah, shipped it to the Grand View area, and modified it to incorporate his new gold-saving ideas.

As the nineteenth century drew to a close, more successful dredge mining technology was applied on Grasshopper Creek in Montana and on the Feather River in California—advances some had anticipated as the coming of a golden age of dredging, not only for western America but also elsewhere in the world.

Contemporary engineers compared progress in applying the new techniques on the Snake River placers to that in New Zealand earlier, with trial and error, repeated failures, and a period of revolution necessary in both cases. On the Snake, one problem was that rich pay streaks were dispersed and the lower gravels were barren, but the major drawback was the fineness of the gold and sometimes its combination with iron, which created special complications for recovering and separating it. In late 1896 one observer noted that all of the hydraulic efforts to raise gravel had failed, "as the mills and dismantled boats and scow tied up along the river will testify." Four years later, in 1900, according to another onlooker, of the large number of dredges operating on the Snake at one time, "only five were still working, but new start-ups were anticipated in the near future."[81] The boosters disagreed. One had noted a few months earlier: "How many dredges are now at work along the river is impossible to state, though every few miles the smoke from one or more may be observed, and in nearly all instance[s], even when not saving a large percentage of the values, these are operating to a profit."[82] At about the same time, an optimistic paper presented at the Washington meeting of the American Institute of Mining Engineers by past president John Birkinbine reported that locals had informed him that sixty dredges were at work on the river—an over-inflated estimate, to say the least. Birkinbine further expressed the hope that low-cost electricity near American Falls could in the future be transmitted some distance to other dredges.[83]

If gold dredging blossomed in the coming years elsewhere in the West, indeed, in the world, its successful technology continued to bypass the Snake River. As the director of the US Mint wrote in his annual report on the production of precious metals for 1902 regarding "a new pattern of dredge" that would save all the fine gold: "That has been claimed for every other dredge, but the public is always glad to believe that the next one will accomplish what it aimed at."[84] New companies and new machines promised incredible results and fantastic profits that never materialized. On the Snake, one after another, dredges appeared, then quickly disappeared, usually leaving stark monuments to hopes never realized.

Notes

1. Merle W. Wells, *Gold Camps and Silver Cities*, Bureau of Mines and Geology Bulletin 22, 2nd ed. (Moscow: Idaho Department of Lands, 1983), 88–91.

2. Ibid., 91–92; *Mining and Scientific Press* 68 (January 27, 1894): 54; Thomas H. Hite, "Fine Gold and Platinum of Snake River, Idaho," *Economic Geology* 28 (May 1933): 256.

3. See Robert N. Bell, "The Origin of the Fine Gold of the Snake," *Pocatello Tribune*, February 15, 1902.

4. Lawrence L. Dee, "Snake River Gold," http://imnh.isu.edu/digitalatlas/geog/historic/histtxt/gold.pdf.

5. *Engineering and Mining Journal* 70 (October 6, 1900): 395; *Engineering and Mining Journal* 73 (February 15, 1902): 241.

6. Don Maguire, "Snake River Gold Fields," *Mines and Minerals* 20 (September 1899): 56–58.

7. Eva Emery Dye, *Stories of Oregon* (San Francisco: Whitaker and Ray, 1900), 162; Eva Emery Dye, *Boise* (Boise: Boise Commercial Club, 1907), 25.

8. This was the Golden Bar Steam Dredging Company, incorporated in Cheyenne in 1895. It had control of 1,000 acres of gravel bars along the river, was said to have tested the ground, and planned to set two large dredges to work early the following spring. No more was heard of the firm. See *The Irrigation Age* (Chicago), January 9, 1896, 53; *Mining and Scientific Press* 71 (November 16, 1895): 319.

9. *Idaho Daily Statesman* (Boise), December 1, 1881.

10. John F. Sanders, "Vacuum Dredge," patent no. 243,462, June 28, 1881, Google Patents; John F. Sanders, "Ore Concentrator," patent no. 258,131, May 16, 1882, Google Patents; *Salt Lake Tribune*, December 1881.

11. John F. Sanders, "Composition for Dissolving the Coating of Gold Ore," patent no. 244,050, July 12, 1881, Google Patents.

12. *Salt Lake Tribune*, January 1, 1882.

13. *Mining Review* (Chicago), quoted in *Ketchum (ID) Keystone*, June 19, 1882.

14. *Salt Lake Mining Review* 4 (December 30, 1902): 35.

15. Ibid.

16. Quoted in A. Wells Robinson, "Recent Improvements in Dredging Machinery," in *Transactions of the American Society of Mechanical Engineers* (New York: Society of Mechanical Engineers, 1886), 3:126.

17. Ibid., 126–27.

18. Ibid., 127. The fact that two companies were involved in both this debacle and the Goose Creek fiasco leads one to wonder if they involved the same group of entrepreneurs or were somehow otherwise connected.

19. *Wood River Times* (Hailey, ID), June 10, 1885.

20. Ibid., July 2, 1890. I have been unable to find any subsequent mention of such a case.

21. See, for example, ibid., May 23, 1883; November 11, 1884; September 8, 1885.

22. *Mining and Scientific Press* 68 (January 13, 1894): 21 (first quote); *Mining and Scientific Press* 68 (January 27, 1894): 54.

23. Ibid. (January 13, 1894): 21; ibid. (May 12, 1894): 292; ibid. 69 (October 27, 1894): 261; *American Engineer and Railroad Journal* 68 (February 1894): 58 ("The gravel" quote); *Ogden Standard Examiner*, November 26, 1893.

24. *Engineering and Mining Journal* 57 (May 19, 1894): 469; *Mining and Scientific Press* 70 (January 12, 1895): 26; *Salt Lake Tribune*, August 18, 1894.

25. *Mining and Scientific Press* 69 (September 29, 1894): 202.

26. Ibid. 71 (August 10, 1895): 87.

27. *Salt Lake Tribune*, April 7, 1895.

28. *Engineering and Mining Journal* 63 (June 12, 1897): 610; *Engineering and Mining Journal* 73 (February 15, 1902): 341; *Idaho Daily Statesman* (Boise), March 4, 1898; *Mining and Scientific Press* 94 (April 27, 1907): 542.

29. Larry R. Jones and Merle W. Wells, "W. W. Priestly's Remarkable Invention," *Idaho Yesterdays* 20 (Summer 1976): 13–14.

30. *Shoshone Journal*, June 26, 1896, quoted in ibid., 14–15.

31. Jones and Wells, "W. W. Priestly's Remarkable Invention," 15.

32. *Mining and Scientific Press* 72 (January 4, 1896): 11; *Mining and Scientific Press* 72 (January 11, 1896): 31.

33. Ibid. 73 (July 23, 1898): 86.

34. *Salt Lake Tribune*, April 10, 1896.

35. *Mining and Scientific Press* 75 (August 28, 1897): 199.

36. Ibid. 72 (February 29, 1896): 171; ibid. (April 11, 1896): 301; ibid. (May 30, 1896): 447.

37. *Idaho Daily Statesman* (Boise), February 25, 1898.

38. Ibid., May 8, 1899.

39. A. V. Williams Jackson and Robert Peele, eds., *A History of the Class of Eighteen Hundred and Eighty-Three of Columbia College* (New York: Irving, 1911), 146–47.

40. *Idaho Daily Statesman* (Boise), August 15, 1891; December 16, 1894; July 26, 1897; November 8, 1897: January 1, 1898; August 30, 1899.

41. *Swan Falls Dam*, Idaho State Historical Society Reference Series 68 (March 1973); *A Historical, Descriptive and Commercial Directory of Owyhee County, Idaho* (Silver City, ID: Press of the Owyhee Avalanche, 1898), 78.

42. *Idaho Daily Statesman* (Boise), August 23, 1898.

43. Richard Z. Johnson, *An Illustrated History of the State of Idaho* (Chicago: Lewis, 1899), 245–46.

44. *Idaho Daily Statesman* (Boise), December 11, 1900; July 28, 1901; *Appletons' Annual Cyclopaedia and Register of Important Events of the Year* (New York: D. Appleton, 1903), 7:714; *Salt Lake Herald*, December 14, 1900.

45. *Idaho Daily Statesman* (Boise), June 9, 1903 (quotes); *Deseret News* (Salt Lake City), June 11 and July 28, 1903; *Salt Lake Mining Review* 5 (July 30, 1903): 4; *Salt Lake Mining Review* 5 (October 15, 1903): 9; *Salt Lake Herald*, November 14, 1904.

46. *Salt Lake Mining Review* 5 (October 15, 1903): 4; *Salt Lake Telegram*, August 28, 1903; *Salt Lake Herald*, July 31, 1905.

47. See United States Patent Office, "Apparatus for Placer-Mining," patent no. 796,780 (August 8, 1905), and "Jet Nozzle for Wet Concentration," patent no. 791,387 (May 30, 1906), both issued to J. L. Weaver, Google Patents.

48. *Idaho Daily Statesman* (Boise), March 12, 1906; April 19, 1911 (quote); *Salt Lake Mining Review* 8 (June 15, 1906): 20; *Mining Reporter* 54 (November 1, 1906): 438–39; C. C. Longridge, *Gold Dredging*, 2nd ed. (London: Cecil Clement, 1907), 285.

49. *Mining Reporter* 53 (June 7, 1906): 571; *Mineral Resources of the United States, 1906* (Washington, DC: Government Printing Office, 1908), part 1, 224; "*Mineau v. Imperial Dredge and Exploration Company*," *Idaho Reports* 19 (March 1911): 459–60.

50. Benjamin A. Dare, "Property of Eureka, Lourdes and Utility Power Companies," typescript, New York, 1907, Fred L. Morris MSS, Bancroft Library, University of California, Berkeley.

51. *Idaho Daily Statesman* (Boise), June 24, 1906.

52. Ibid., April 21, May 29, August 19, 1898.

53. Ibid., January 1, 1902.

54. Ibid., April 8, 1906. These dredges were probably built by Isaac Hammond's firm. Before he moved west, Hammond had lived in Kankakee, Illinois, for a time.

55. *Deseret Evening News* (Salt Lake City), January 24, 1900.

56. *Spokane Chronicle*, quoted in *Salt Lake Tribune*, July 12, 1899.

57. Ibid.

58. *Salt Lake Herald*, May 28, 1900.

59. Ibid.

60. An excellent brief biography of Tracy is found in the description of the Tracy papers, on deposit in Boise State University's Albertsons Library. See http://boisestate.worldcat.org/title/paul-e-tracy-papers-1875-1976/oclc/29310119&referer=brief_results.

61. From Paul E. Tracy, "The Dredge Tender," n.d., copy, typescript, Special Collections and Archives, Albertsons Library, Boise State University, Boise, ID, courtesy of Lorna Tracy.

62. Ibid.

63. Frederick Powell, *Engineering and Mining Journal* 70 (October 6, 1899): 395–96; Robert Brewster Stanton, *The Hoskaninni Papers: Mining in Glen Canyon, 1897–1902*, edited by C. Gregory Crampton and Dwight L. Smith, University of Utah Anthropological Papers 54, Glen Canyon Series 15 (Salt Lake City: University of Utah, 1961), 92.

64. *Idaho Daily Statesman* (Boise), November 16, 1898; February 12, 1902; *Salt Lake Mining Review* 2 (April 15, 1901): 25.

65. *The Engineering Record, Building Record and Sanitary Engineer* 43 (March 9, 1901): 239.

66. *Idaho Daily Statesman* (Boise), February 12, 1902 (quote); *Mineral Resources*, 604.

67. *Salt Lake Mining Review* 3 (August 15, 1902): 23.

68. *Salt Lake Telegram*, April 17, 1902.

69. J. M. Hill, *Notes on the Fine Gold Snake River, Idaho*, US Geological Survey Bulletin 620-L (Washington, DC: Government Printing Office, 1915), 277.

70. *Idaho Daily Statesman* (Boise), April 7, 1901; *Deseret News* (Salt Lake City), April 11, 1901.

71. *Idaho Daily Statesman* (Boise), May 19 and July 17, 1903. *Mineral Resources 1914* lists "Idaho Gold" as active in Bingham County for 1902 (604). This is probably a mistake on the compiler's part. The company appears to have dug only in 1903.

72. *Report of the Mining Districts of Idaho for the Year 1904* (Boise: Statesman Shop, n.d.), 32–33, cited in *Idaho Daily Statesman* (Boise), May 19 and July 17, 1903; *Eighth Annual Report of the Mineral Industry of Idaho for the Year 1906* (n.p., n.d.), 22; Alfred R. Schultz, "Gold Developments in Central Uinta County, Wyo., and at Other Points on the Snake River," US Geological Survey Bulletin 315 (Washington, DC: Government Printing Office, 1907), 80.

73. *Idaho Daily Statesman* (Boise), April 7, 1901.

74. *Salt Lake Mining Review* 2 (December 25, 1900): 12.

75. Ibid.

76. Ibid. (January 15, 1901): 21.

77. Ibid. 3 (July 30, 1901): 21.

78. *Salt Lake Mining Review* 2 (July 30, 1900): 11; *Salt Lake Mining Review* 3 (August 15, 1901): 18; *Salt Lake Herald*, March 29 and April 14, 1901; *Idaho Daily Statesman* (Boise), October 21, 1901 (quote); *Deseret Evening News* (Salt Lake City), November 18, 1901.

79. Bucyrus Company, "List of Dredges," 100075A–100075D, south Milwaukee, n.d., copy in possession of the author.

80. *Idaho Daily Statesman* (Boise), April 8 and November 16, 1906; *Salt Lake Telegram*, June 17, 1902; Crampton and Smith, *Hoskaninni Papers*, 92. Working from a handwritten manuscript, Crampton and Smith erroneously refer to the "Golden Grant Dredging Co."

81. William W. Washburn, *Mining and Scientific Press* 73 (December 26, 1896): 522; Frederick Powell, "Gold Dredging on the Snake River, in Idaho," *Engineering and Mining Journal* 70 (October 6, 1900): 396.

82. *Salt Lake Herald*, May 28, 1900.

83. John Birkinbine, "Hydraulic Pumping-Plant on the Snake River, Idaho, for Power, Irrigation and the Treatment of Gold Sands," *Transactions of the American Institute of Mining Engineers* 30 (February–September 1900): 522.

84. J. W. Cunningham, *Report of the Director of the Mint upon the Production of the Precious Metal in the United States during the Calendar Year 1902* (Washington, DC: Government Printing Office, 1903), 141.

2

The Black Sands Craze

"Process mania" had long been an integral part of American mining history and had contributed to many companies' losses of investors' funds, despite constant admonitions to gullible investors. Fantastic but ill-conceived "sure-fire" equipment and worthless procedures to save and separate precious metal were as commonplace in placer operations as in hard-rock mining. One early writer commented disdainfully on the "improved-back-action-lightning-gum-elastic-cylinder-and-Spanish-fly amalgamator," with which the self-styled "expert" could "*draw* gold from a Rocky Mountain turnip." Another warned that "the new process fiends are here, and are as plentiful as mosquitos in a Canadian swamp." Beware of the "'new process,' which in its working will completely revolutionize hitherto known to chemical and physical science." As for new machines, "Our inventor knows as little about practical mining as the lunatic in Swift did about extracting sunbeams from cucumbers."[1]

Gold dredging had fewer pitfalls of this sort than other types of mining, but it was not immune—especially in the early years when the industry was

DOI: 10.5876/9781607324751.c002

feeling its way. Witness the Goose Creek Ferry fiasco in 1882, followed by numerous experimental and failed dredge machines on the Snake. That was the way of the trial-and-error approach. Concerned by the increasing scarcity of the world's supply of platinum, in early 1905 the US Congress directed the US Geological Survey to include in its annual report on mineral resources an investigation of the useful minerals contained in the black sands of the Pacific Slope—a charge subsequently expanded to embrace the entire country. Survey scientist David T. Day and Robert H. Richards, head of the mining department at Massachusetts Institute of Technology, directed the project. They sent out a call for samples to more than 8,000 known placer miners, including a number in Idaho. Victor Hikes of the US Geological Survey made arrangements for shipping samples from various points on the Snake; Captain Relf Bledsoe collected twenty sacks of gravel and sand from Big Foot bar, a dozen mines above the Swan Falls power plant; while a group headed first by L. G. Gillette and then by W. L. Walker, both owners of placer ground, gathered gravel from 200 excavations along the Snake between Minidoka dam and a site opposite Wapi.[2]

Most samples were heavy waste sands discarded from sluice boxes in the cleanups of placer mining, including dredging. If these concentrates contained gold or platinum, it was in such a combination that the miner was unable to save either of them by any familiar process. The sands were of six kinds: sea sand, river sand, hydraulic tailings, gold dredge tailings, placer cleanup sands, and gold dredge cleanup sands. A small army of assayers, metallurgists, and other mining experts was enlisted to analyze and break down the samples into their component parts, using a variety of machines in their work. Results showed at least sixteen "salable" minerals in the black sands, including gold and silver, and Day and Richards noted the seemingly endless quantities of them on the shores of oceans, lakes, and rivers in the West. They invited innovators to develop new machinery and processes to handle the black sands more effectively and cheaply. They also demonstrated on the scene that magnetite, a common iron oxide, could produce pig iron and steel in a small electric furnace. But because of the cheapness of iron ore, the men concluded that it did not fall into the category of a "useful mineral."[3]

Publication of their report touched off renewed interest not only in the Snake River placers but also in new methods of handling and processing the refractive sands. One sample, from west of Blackfoot, for example, showed 19.62 ounces per ton of concentrates in gold and 0.18 in platinum, as well as other rarer metals and enough iron that the Boise *Statesman* noted that "Idaho is a sharer in the now celebrated steel-in-black-sands discoveries of Dr. David T. Day." "Big Fortune in Black Sand," shouted the newspaper's headlines.[4] Reporter Frank Carpenter's interview of Day received a half-page spread in the *Deseret News* and was no less effusive: "Snake Exceptionally Rich" and "Will Add Millions of Our National Wealth." What David Day might have said but didn't, according to Carpenter, could have been, "I am like King Saul the son of Kish, I started out to hunt my father's asses and lo 'I found a kingdom.' "[5]

"The combined gold of the black sand concentrates of the Snake were now the "largest and richest natural sluice box in the world, and perhaps the greatest deposit of gold in the world," wrote George W. McGhee, a metallurgist who had studied the region for three years and published his findings in *Ores and Metals.*[6] He had obtained these concentrates from river bars at depths of one or two feet to as deep as ten or twelve feet below the gravel and at different points over a distance of 240 miles so he could treat and analyze hundreds of samples to determine their gold values. He estimated that the free, flaky metallic gold on the gravel bars had an average worth of around ten cents per cubic yard of gravel. Thus, since only a small percentage of the value was gained by amalgamation, the more valuable black sands were dumped back into the water. In treating those sands from one or two feet below gravel level, McGhee estimated that gold values ran from $400 to $700 a ton, including both free and combined gold. At Glenns Ferry, from three feet down, values ran at $1,300, while on the grounds of Imperial Dredge and Exploration Company at Grand View, values were $1,048.80 per ton. Farther west, a forty-pound sample taken by a suction dredge at Great Bend from ten feet to twelve feet down showed gold values of $8,270 a ton.[7] McGhee put the estimated cost per yard of obtaining the concentrates at from $2.50 to $3.00 per cubic yard using a sand pump or a ten-inch suction dredge.[8] These computations were far in excess of the much more modest and conservative findings of the Day-Roberts report. Ignored was the fact that the samples had been castoff concentrates, the results of the sifting of countless tons of sand and gravel; that river bars were often fickle where paying deposits were concerned, and that commercially viable processes did not yet actually exist.

This was heady stuff, the more so when "experts" offered secret surefire procedures for separating the metallic values from the stubborn black sands. Investors' hopes in failed river-dredging ventures were regenerated, as a flurry of optimism was followed by mining men and scientists developing processes tailored to the new revelations.

The 1905 *Mineral Resources of the United States* believed the Portland tests to some extent stimulated the building of dredges and updating the equipment of outdated boats. In addition, they strengthened the resolve of operations already under consideration along the Snake—specifically those below Blackfoot, American Falls, Glenns Ferry, and Weiser.[9] In the spring of the following year, a "well-known mining engineer," who wanted his name withheld, made the rounds of western Idaho camps and concluded that there was "unusual activity" along the Snake River at these same locations.[10]

Particularly intriguing were Day's experiments with magnetic iron sands and onsite production of pig iron and steel with an electric furnace. Actually, the process was not a commercial success, although it was much discussed and considered a technological breakthrough by the public. One hydraulic and placer expert noted that "to spend $5 to recover $1 is not a scientific discovery, and the whole affair was the cause of much amusement to mining engineers and metallurgists."[11] Dreamers on the Pacific Coast could discuss construction of "battleships from sand" for the future.[12] Steel shipped from the East added another ten dollars a ton to the cost, which could be eliminated by establishing plants to handle black sands in the West. A number of abortive efforts were made to do just that.

Some of them were what one mining journal called "the efforts of so-called scientists and inventors to separate gold from the non-technical public by alleged mechanisms and processes, the latter generally secret." These charlatans made much of the Day report, the accuracy of which no one doubted but which nowhere imparted commercial values to the experiments or implied that the deposits could produce practical metallurgical profits.[13] Black sand mining offered a fruitful field for unscrupulous promoters, some of whom claimed they could recover gold and platinum even if they could not be detected by any known scientific method. Countless wrecks of experimental machines attested to "the credulity and childlike simplicity of those who have been induced to invest their money in schemes for treating the sands."[14]

In contrast, the investigation was also followed by legitimate but generally futile efforts to work the sands. In the spring of 1907 James H. Brady, president of the American Falls Power, Light, and Water Company, announced that he would build a plant to treat ores from northern and southern Idaho and from Utah and that he would "convert the black sands of the Snake river into precious metals." Orders for machinery had already been placed in Schenectady, New York, and Brady hoped to have a small plant operating within ninety days at American Falls. He mentioned no particular process but noted that electrical treatment of ores had been tested successfully in France, Germany, and Canada and that he had done intensive investigation of the subject. He saw a bright future ahead: "I have no doubt that the installation of a small plant at American Falls will result in a few years in building up a big industry there."[15] There is no evidence that such a venture was ever completed.

In the spring of 1912 the Calidaho Extradition Company announced plans to erect a ten-ton plant in Twin Falls to treat black sand and recover its gold by an electrochemical process.[16] Heretofore, said spokesmen for this San Francisco company, "All up and down the Snake river, especially in the canyon in this vicinity, old machinery, tunnels and ruins of cabins tell the story of 50 years of placer mining and failures." Now this industry "is to be revolutionized and given new impetus because of a new process just recently invented."[17] Familiar words indeed. The new treatment promised to extract more gold from black sand than had any other mode of handling—60 percent more than the regular smelting procedure. It also amalgamated platinum and other metals. Calidaho Extradition held exclusive patent rights to the new process for use in Idaho and parts of Oregon, and its president stated that the Twin Falls works was to be the prototype of a number of such plants designed to process from ten tons to fifty tons of black sand per day.[18] Again, nothing appears to have happened.

Back in the original Sweetser-Burroughs country, the black sand report touched off one belated outburst of interest from the Empire State Dredging Company, a firm that also did harbor and river dredging. It was financed by a group of New York capitalists headed by William J. Van Vleck—a dentist by trade—and was said to have extensive holdings of mining property in Oregon, Montana, and Canada.[19] Construction of the Milner dam in 1905 and the new Minidoka dam the following year not only opened substantial acreage for irrigated farming, it also offered new placer possibilities.

Prior to completion of the Minidoka dam in 1906, Diamond bar on the Cassia County side of the river was deemed potentially rich but almost impossible to handle with a dredge. In addition to the usual problems of saving the fines and finding an economical means of breaking down the refractory black sands, the ground was too high and too porous to get water on it to make it workable. With the help of the dam, the US Geological Survey, and its own laboratory research, Empire State now believed solutions to these difficulties were at hand.

To begin with, the dam covered the bar with from six inches to thirty feet of water, thus making it a viable dredging proposition. At the same time, sophisticated analyses and testing of black sands at sites around the country by US Geological Survey scientists showed remarkable values not only in gold and platinum but also in magnetic iron in quantities that would justify exploiting the sands alone.[20] While Diamond bar was not included in the sampling, the company conducted its own research and found much to like by way of flake gold and black sands on its own grounds. Moreover, officials claimed to have perfected a process for handling the recalcitrant black sands that would "work the same sort of revolution in placer mining as has the cyanide process in the handling [of] refractory quartz ores." Successful breakdown of the sands into elemental parts of gold, platinum, iron, tungsten, and manganese was based on a secret chemical treatment that promised to save about five times as much of the precious metals as any other known process.[21]

In early 1907 General Manager William H. Godfrey of Butte was in Minidoka to make the second payment on the purchase of 2,000 acres of the Lilly placers on the bar. Although delayed somewhat by ice on the river, the company had begun work on building a separate concentrating plant onshore. It would start with the twelve-year-old Sweetser-Burroughs boat, on which new centrifugal suction machinery had been installed, but it planned to install two massive twelve-foot bucket dredges later. If sufficient iron and manganese were produced, there was even the possibility that a steel mill might be built onsite.[22] Despite the eternal optimism that pervaded the mining world, the Empire State Dredging Company's dredge efforts seem to have been fruitless on the Snake: as in so many other cases, nothing happened.

Another highly publicized example of process mania that seemed to run the length of the Snake in Idaho was the product of "two noted eastern scientists," Dr. Oscar Eckstein and Professor Wilfred G. Fralick. Fralick was a New York physician—"one of the most celebrated surgeons in the United States," according to one editor in 1902, who had been appointed a professor at the University of Chicago.[23] The press reported that he had discovered "a universal germicide" expected to be a panacea for all kinds of infectious diseases, especially tuberculosis, although Fralick's own hospital disavowed any connection with the doctor's alleged cure for that particular malady. The German-born and educated Eckstein was a chemistry professor at the University of Chicago, where the two men had worked together for several years solving scientific problems.[24]

According to Fralick, Eckstein's special expertise was working the black sands for their gold content. His method of extraction was chemical, was simpler than the standard amalgamation or cyanide approach, and promised to save as much as 95 percent of the values, far more any other process.[25]

In the summer of 1906, the two were in Idaho. Fralick was on a brief holiday. While scouting out mining possibilities on the west side of the Snake near Farewell Bend and Old's Ferry, both men had unexpectedly found ground along the river that looked extremely promising and had quietly filed claims on it, much to the consternation of squatters who for years had been "scratching together a meager living, unconscious that a fortune lay in the gravel on which they walked." Eckstein doubted that his colleague would return to his medical work immediately. "Dr. Fralick is suffering from an attack of the gold microbe," he said, "and may protract a vacation longer than he had at first intended."[26]

The two scientists—now entrepreneurs and mining men—announced that a test plant would be installed on their property at once but that it might be a year before the facility would be ready to handle the reduction of black sands from all comers. Then, having staked a third claim, all of the sixty-acre island at Farewell Bend, they arranged for eastern engineers to draft plans for the plant and invited every placer operator in the vicinity, whether working with the smallest hand process or the most extensive dredging method, to patronize the facility. In the meantime, they were negotiating with some of the largest dredge operators in Idaho, California, Wyoming, and Montana and planned with great secrecy to build reduction plants in Alaska, Australia, South Africa, and anywhere else black sands were abundant.[27]

Several dredging companies displayed an interest in the process, and at least one, the Idaho Placer Mining Company, sought to utilize it. That firm was founded by a former ship captain turned entrepreneur, Charles Green, who came up with a new piece of equipment that was going to save the fines and revolutionize the placer industry. In the summer of 1904, according to the Weiser newspaper, Green was promoting a new dredge endeavor that would take care of all previous difficulties. Backed by Boise capitalists as well as local investors, he organized the Idaho Placer Mining Company, which acquired 200 acres along the Snake at Old's Ferry, just west of Weiser. Green was president and general manager of the firm. With $10,000 from the company, Green arranged to build a large double-decker flatboat. On the lower level would be the latest type of dredging machinery, while the upper level would be used for living and storage. Fifty feet long and thirty feet wide, the boat would be equipped with a 100 horsepower boiler and a 20 horsepower engine to run a patented suction apparatus to draw the sands through a twelve-inch hose into a hopper, where the material would pass onto a grizzly or rocking screens and the gold would be separated. The cost of the rig was estimated at about $8,000, and it would be able to handle 3,000 yards per day, at an expense of $60 per twenty-four hours, or roughly 2 cents a yard.[28]

Apparently, this venture was slow in coming to fruition. During the winter, Green consulted with builders of mining machinery in Portland and Seattle, as well as in the East, where he sought additional funding. In the spring of 1905 he returned with financial backing and carefully tweaked plans for the ultimate dredging machine. In addition to saving the fines, Green also expected the machine to recover the black sand gold through special smelting. The first carload of machinery was expected at Old's Ferry "at an early day," after which construction would begin immediately.[29] In early 1906, in a corporate reorganization, the Idaho Placer Mining Company added the word "Limited" to its name, gained new investors from Chicago, and prepared to commence operating "at once."[30] Whether it did so is not clear. In the interim, the new company sued Green. It insisted that he had only spent $6,000 of his allotted $10,000 and demanded a return of the remaining $4,000. Green contended that had spent all of the money plus $1,000 of his own; in addition, the Idaho Placer Mining Company, Ltd., owed him $1,800 in salary for superintending the construction of the dredge.[31]

In late September 1906 J. L. Kline was back home in Boise from the Old's Ferry area, where he had been doing work for the company preliminary to the installation of machinery at Farewell Bend for the extraction of gold from black sand. The machinery was expected to arrive in about ten days.[32] Very likely, the dredge firm had signed on with the Eckstein-Fralick processing group.

But nothing seems to have happened for nearly two years, at which time the Idaho Placer Mining Company announced a projected startup on about September 1, 1908, a date that was delayed for several more months. It also announced that it was constructing an "experimentation station," at a cost of $30,000, to utilize a Chicago chemist's secret process for treating black sands. The names Eckstein-Fralick were not mentioned, but no doubt they were involved. General Manager E. C. Cleaver noted that the firm's suction dredge was specially adapted to the task of concentrating the black sands prior to the chemical treatment thereafter. The experimental station would include accommodations for workers, beginning with ten, with the number increasing as work progressed. If the company was successful, another dredge would be added.[33] Ambitious plans, these, but only a small part was fulfilled.

The boat did some digging in 1909, and it was reported that black sand from Old's Ferry was being shipped to New York City for the new chemical treatment. Because of "the unsettled condition of affairs"—whatever that may have meant—the big rig had remained idle for some time, but in the spring of 1911 it was operating again under the watchful eye of Superintendent Eastman, "with satisfactory results." Black sands were now shipped to what was called "the company's big reduction works in Chicago" to undergo both electrical and chemical treatment.[34] How long it was in operation is a mystery, but it was probably not long.

In late 1903 the Boise *Statesman* reported that the Canton–Snake River Gold Company was building a dredge at Glenns Ferry to work on the river. In addition, it had ordered a twenty-ton steamboat from Portland to carry freight and passengers regularly between that town and Bruneau. Fifty feet long and drawing only eight inches of water, the boat was said to be able to carry sixty tons per trip, and officials of the dredge company planned free excursions later to introduce the service. At the time, the company's two leaders were County Auditor Daniel W. Shetler of Mountain Home and Roy Raber of Canton, Ohio.[35] The bucket-line dredge was powered by two large marine boilers and four huge steam engines and was expected to dig up to 7,200 cubic yards a day. Moreover, the firm had contracted to use N. L. Raber's new secret electrical concentration and amalgamation process exclusively, touted as "an established success," a description that created skepticism in the minds of at least some Idahoans. The *Idaho City World,* for example, scoffed at the news that Roy Raber had taken a container of mixed flour gold and sand and that, after half an hour's work, had come out with the gold in one bottle and the sand in the other. "That looks too much like legerdemain," said the *World* editor, who pointed out that in the past there had been so many "demonstrated successes" of new ways to save fine gold that all processes were now looked upon as almost certainly, "paradoxically speaking, successful failures."[36]

These efforts seem to have come to naught. Then came the Day-Richards report. In early 1906 the Ida-Elmore Company announced that it was ready to start up the old Canton machine above Glenns Ferry.[37] The dredge had been lying idle for some time because of a breakdown. Daniel Shetler and a Mr. Spranger, the latter representing Canton interests, painted a rosy picture. The equipment was solidly built and saved a higher percent of precious metal than any other boat operating on the river. The steamer *Helen* would be ready to tow coal and other supplies. Once everything was shipshape again, the spokesmen expected "a prosperous and busy season of placer mining."[38] Alas, this was but another of the Snake's "successful failures." Again, the results did not lead to any sustained digging.

Later that same year, J. N. Parks, the treasurer of the Idaho Gold Extraction Company, was operating a dredge on the Snake near Glenns Ferry. Although the dredge had been working for only three weeks, Parks believed the results demonstrated the richness of the ground and warranted building two boats for the following year. Because of low water, the dredge was digging the river bottom rather than the company's extensive placer gravels onshore. Parks and W. J. Witham, the company's manager, went to Boise to negotiate with Dr. Fralick for use of his exclusive process.[39] No further information was forthcoming.

Here the Eckstein-Fralick story ends—not with a bang or even a whimper. Like so many of the Snake River dredging schemes, this one simply fades away. In all probability, the two men were not dishonest; they likely truly believed in their process and sought to provide a service to the placer industry, at the same time lining their own pockets. Either the process itself was flawed, or the black sands were not concentrated enough to make their operation economically successful.

FIGURE 2.1. This photo shows the beginning of a transfer of an early dredge from the Snake River near Glenns Ferry to a new site ninety miles away, not an unusual event. Courtesy, Idaho State Historical Society, Boise, ISHS 72-112.b.

In 1938 the US Bureau of Mines put out an information circular in which it warned individuals and companies against being too optimistic regarding the chances of striking it rich by handling black sands. If there were profits to made in them in later years, it was not from the retrieval of gold but rather from the rare minerals therein—many of them now considered vital to modern society.[40]

Notes

1. T. O. Bigney, ed., *The Rocky Mountain Directory and Colorado Gazetteer, for 1871* (Denver: S. S. Wallihan, 1870), 245; *Engineering and Mining Journal* 60 (November 16, 1895): 462; *Colorado Mining Journal* (Gunnison) 1 (September 16, 1884): 210; Amasa McCoy, *Mines and Mining in Colorado: A Conversational Lecture, Delivered in the Lecture Room of Crosby's Opera House, to the International Mining and Exchange Company* (Chicago: Printed privately for the company, 1871), 35.

2. *Idaho Daily Statesman* (Boise), July 26 and August 17, 1905; *Deseret News* (Salt Lake City), August 19, 1905; US Geological Survey, David T. Day and Robert H. Richards, "Useful Minerals in the Black Sands of the Pacific Slope," in *Mineral Resources, 1905* (Washington, DC: Government Printing Office), 1178.

3. Day and Richards, "Useful Minerals," 1175, 1176–77, 1179, 1228–29.

4. *Idaho Daily Statesman* (Boise), December 1, 1905, April 8 and September 3, 1906.

5. *Deseret News* (Salt Lake City), August 21, 1906.

6. George W. McGhee, "The Black Sands of the Snake River, Idaho," *Ores and Metals* (Denver) 14 (August 20, 1907): 313.

7. Ibid.

8. Ibid., 314.

9. Day and Richards, "Useful Minerals," 219.

10. *Deseret News* (Salt Lake City), March 24, 1906.

11. Eugene B. Wilson, *Hydraulic and Placer Mining*, 3rd ed. (New York: John Wiley and Sons, 1918), 326.

12. *Electrochemical and Metallurgical Industry* (New York) 4 (December 1906): 474.

13. Ibid. 27 (July 1907): 547.

14. *Chemical Engineer* (New York) 27 (February 1919): 34.

15. *Idaho Daily Statesman* (Boise), April 19, 1907.

16. *Mining and Scientific Press* 104 (March 2, 1912): 355.

17. *Idaho Daily Statesman* (Boise), February 19, 1912.

18. Ibid.

19. Ibid., November 26, 1906.

20. See David T. Day and Robert H. Richards, "Investigation of Black Sands from Placer Mines," in *Contributions to Economic Geology 1905*, USGS Bulletin 285, Series A, vol. 73 (Washington, DC: Government Printing Office, 1906), 15–165.

21. *Idaho Daily Statesman* (Boise), November 26, 1906.

22. Ibid., February 5, February 8, April 7, 1907.

23. *Arizona Daily Journal-Miner* (Prescott), July 19, 1902.

24. *Idaho Daily Statesman* (Boise), September 3, 1906; *Medical Standard* (Chicago) 24 (November 1901): 629; *American Medicine* (Philadelphia) 2 (December 1901): 889.

25. *Idaho Daily Statesman* (Boise), August 27, 1906.

26. *Electrochemical and Metallurgical Industry* (New York) 4 (December 1906): 474.

27. *Idaho Daily Statesman* (Boise), September 4, 16, 17, 1906.

28. Ibid., June 23 and 28, 1904; *"Idaho Placer Mining Company v. Green," Pacific Reporter* 94 (1908): 161–62.

29. *Idaho Daily Statesman* (Boise), April 21, 1905.

30. Ibid., March 25, 1906; *Mining and Scientific Press* 92 (March 31, 1906): 228.

31. *"Idaho Placer Mining Company v. Green,"* 161, 163.

32. *Idaho Daily Statesman* (Boise), September 28, 1906.

33. Ibid., August 21 and December 7, 1908; *Salt Lake Mining Review* 10 (December 15, 1908): 19.

34. *Idaho Daily Statesman* (Boise), April 19, 1911; *Salt Lake Mining Review* 13 (May 15, 1911): 21.

35. *Idaho Daily Statesman* (Boise), November 6, 1903; *Salt Lake Mining Review* 5 (August 15, 1903): 17.

36. *Salt Lake Mining Review* 5 (March 30, 1904): 6; *Salt Lake Telegraph*, October 5, 1903.

37. *Mining and Scientific Press* 92 (February 17, 1906): 132.

38. Glenns Ferry, *Signal*, quoted in *Salt Lake Mining Review* 8 (February 28, 1906): 20.

39. *Idaho Daily Statesman* (Boise), October 16, 1906.

40. M. W. Von Bernewitz, "Treatment and Sale of Black Sands," *US Bureau of Mines, Information Circular* 700 (March 1938): 1.

3

The Burroughs Brothers

One of the exceptions to the many rigs that had disappointing results was the Sweetser-Burroughs dredge, which went into the Snake near its convergence with the Raft River thirty miles from the railroad head at Minidoka and was the product of evolution over several years. Lewis Sweetser and the Burroughs brothers had been classmates at Yale University and had graduated from Yale's Sheffield Scientific School with degrees in engineering. George and Harry Burroughs were sons of a well-to-do Chicago distiller, who lost his business in a fire but subsequently prospered in the battery industry as the age of the automobile began. Young Sweetser's father owned large livestock and fruit lands in California and had been one of the pioneering cattlemen in south-central Idaho, with a spread along the Raft River that ran from near present-day Malta thirty miles to where it emptied into the Snake. In the early 1890s the elder Sweetser sold a portion of his ranch to his son and the two Burroughs boys, who proceeded to run a small herd of their own on their "Bar Y" ranch. (The "Y" stood for Yale. Continuing to reflect their old school ties, they also applied for and were awarded a post office for

DOI: 10.5876/9781607324751.c003

the town of Yale.) They couldn't help but be aware of the various small-scale placer operations along the river, and when the cattle market did poorly over the next few years they decided to try their luck in mining as well.[1]

Given their background and training, it is not surprising that these young entrepreneurs looked to more efficient technology. Gold dredges had been effectively applied in New Zealand for many years, and experiments with various dredge equipment had gone on in California, albeit with much less success, for decades. The Snake itself had at least a brief history of mechanical fiascos and short-lived efforts, and the Yale graduates must have been mindful of some of them.

The Sweetser-Burroughs Mining Company was organized in 1893, with George Burroughs Jr. as president and manager. Its first dredge was planned for that autumn. Lumber and machinery were ordered during the winter, probably from Chicago, and as soon as weather permitted in the spring, construction began at a point on the river not far from where the town of Burley would develop.[2] By May 1895 at least, George Burroughs announced that the dredge was beyond the experimental stage "and is now an assured success." The editor of the Albion *Busy Bee* responded with his own forecast: "We predict that within the next two or three years almost hundreds of these same machines will be anchored in the waters of the snake, while their engines will be pumping to the surface thousands of dollars in gold dust annually."[3]

The *Argus*, as the machine was called, was a steam-powered suction dredge, technically described by experts as a hydraulic dredge. According to a contemporary newspaper, "The plant consists of a sand pump and burlap machines set upon a flatboat. The sand and gravel are pumped up from the bottom of the stream and run over the burlaps. The dredge works against the bank so that some of the material comes from the bottom of the river and some from the bank."[4]

The "burlap machines" were an adaptation of equipment originally used in small placer operations by individual miners and now applied on a larger, more mechanized scale by the company. According to George Burroughs, they were able to save at least four-fifths of the gold, probably an optimistic figure.[5]

Since Snake River gold was frequently associated with magnetic iron sand, separation was difficult, not only because of the fineness of the particles but also because some were in the shape of flat scales, more easily lost than the minute grains. Both screening and direct amalgamation were tried but

FIGURE 3.1. The Sweetser-Burroughs Mining Company suction dredge, the *Argus*, was the only dredge on the Snake River that ever paid a dividend. Courtesy, Edgar Rice Burroughs, Inc., © 1975, all rights reserved, TM Edgar Rice Burroughs, owned by Edgar Rice Burroughs, Inc., and used by permission.

rarely proved successful, except on a small scale. If amalgamation using copper plates were to produce a profit on ten-cent gravel, a dredge would have to handle the equivalent of about 1,700 tons of material per day, and the large amount of plate surface required would make the process impractical. Hence the only pragmatic method was the "burlap system," or the "sack box," as the locals called it.[6]

With a hull thirty feet by ninety feet and drawing three feet of water, the *Argus* was a complex machine. At first, a 75 horsepower engine powered its centrifugal pump and lifted the gravel through a thirty-foot pipe with a six-inch nozzle. A year later, the nozzle had been expanded to a ten-inch diameter and then to a foot by the spring of 1897. A second engine drove the elevator mechanism, which carried the tailings and dumped them behind the hull. As the boat moved back and forth across the river, the open end of the suction pipe remained against the gravel deposits, so that everything loose was drawn in—occasionally including rocks weighing as much as fifty pounds—and was carried to a stationary sluice six feet above the floor of the boat. Next came two shaking grizzlies, each seventeen-and-a-half feet long and five feet wide and set at a steep pitch, so arranged that the smaller material was screened out onto burlap-covered tables while the coarse stuff was pushed out by steel scrapers on an endless belt to drop into the river behind the stern. There were sixteen burlap tables—eight on each side—each fifteen feet long and three feet wide, positioned at right angles to the sluices and extending over the sides of the hull. Fine gold and black sand collected in and under the burlap mesh, and when the tables were cleaned, the concentrates were ready for amalgamation. Just as the company quickly dropped any idea of traditional riffles in favor of a large-scale application of "the burlap system" to save the gold, it also soon discarded the idea of copper plate amalgamation, which would have required too much surface space for use on the confines of a dredge, in favor of the simple, time-honored process of amalgamation with mercury. Without interrupting digging operations, the burlaps were taken up at intervals of six or eight hours and washed in a wooden tank, from whence the gold and black sand concentrates were raised automatically to the amalgamating room. There they were combined with mercury in a revolving barrel. The resulting amalgam was then retorted and the gold melted and poured into bars.[7]

The pump was capable of delivering 200 cubic yards of gravel an hour. By mid-1897, after adjusting to the nature of the gold, the spotty deposits, and the river environment, the dredge had been remodeled three times and in the last year had been much enlarged, to the extent that to duplicate it would cost an estimated $15,000 to $20,000. Robert N. Bell, Idaho's inspector of mines, who was close to George Burroughs, said that the original investment was $1,200, "and a whole lot of rare mechanical skill [went] in[to] its evolution."[8]

Initially, the boat ran only during the day, with operating expenses of roughly twenty dollars per diem, but the owners believed that with a dredge force of seven men it could run twenty-four hours a day, and soon three shifts were working round-the-clock. In addition, thirteen men had been hired to bring in fuel for the engines. Some were hauling cedar; some were cutting sagebrush, which made a better fire but cost more.[9] Eventually, Wyoming coal was used to supplement other fuels.

As the suction dredge adequately demonstrated its profitability, the Sweetser-Burroughs company added a second boat—this time a bucket-line dredge of the type catching on in California and elsewhere in the mining West. The *Yale*, as it was called, was built in 1898 by Isaac B. Hammond in Portland, Oregon, and J. W. Burgess of Boise, "a skillful mechanic," was hired to install the machinery in the new hull, a job that would take roughly three weeks. Both boats were expected to run through the coming summer.[10] Equipped with five-cubic-foot buckets, the *Yale* could handle 2,000 yards to 2,500 yards of sand and gravel in a twenty-four-hour day, and over its first twenty months of operation, it employed the same number of men as the *Argus*. It was powered by an 80 horsepower boiler and one horizontal automatic engine that ran all the machinery onboard, including seven winches for various functions, the ten-inch centrifugal pump that provided water for the tables, and the generator for lighting the plant at night. Power for each separate portion of machinery was transmitted by way of a friction clutch operated by a lever in the pilot house. The engine burned 2.5 tons of coal in twenty-four hours.[11] The gold-saving apparatus on the *Yale* was similar to that on the hydraulic *Argus*—a long grizzly and shaking screens for separation, then multiple burlap sluices to catch the values, and finally mercury amalgamation and retorting. One new feature on the bucket dredge was the addition of a small auxiliary pump to throw jets of water to salvage materials clinging to the sides and corners on inverted buckets after they dumped their loads.[12]

The fact that this, too, was largely a family enterprise was indicated by the new Yale Dredging Company, created to handle the new machine. With capital stock of $100,000, the firm's board of directors consisted of Lewis Sweetser, three of the Burroughs brothers—George, Harry, and Coleman—and Walter S. Sparks, a longtime affiliate of the family's ranch and mining operations.[13]

Later in 1898, after a season that was "entirely satisfactory," George Burroughs visited Boise Basin to observe dredge operations there, and the *Statesman* took the occasion to laud him as the "first man to make a success of dredge mining on the Snake."[14] Late in 1899 Burroughs noted that the dredges were getting fifteen cents to twenty cents a yard in the river between Cassia and Lincoln Counties. "This is nothing heavy," he said, "but it is a paying proposition, and there is good dredging in sight for a long time to come."[15]

George Burroughs pointed out that since the spring of 1894, when the dredge started up, only the burlap tables had proved effective; he believed this was the only operation of any size successful in saving the flour gold that ran values of ten cents a cubic yard in that part of the Snake. In this, he was simply stating what other knowledgeable experts had acknowledged: that the basic problem with making placer mining pay on the bars of the Snake was not so much the saving of the fine metal as the ability to process it in sufficient quantity from low-grade gravel to warrant a large capital outlay. Burroughs and Sweetser seemed to have come up with a viable answer. As the *Engineering and Mining Journal* observed a few years later: "The Sweetser-Burroughs Mining Company has set the pace, and under similar competent management the Snake River Valley would seem to offer a field of great possibilities for profitable gold dredging. In the future also the waterfalls and rapids now running to waste along this stream may be taken up and transformed into electric energy."[16]

Without a doubt, George Burroughs was an exceptional manager with an enlightened policy toward his workers. Once established, the workforce of the *Argus*, the first boat, consisted of three men per eight-hour shift, including a chief engineer, a blacksmith, and roustabouts.[17] Describing the new *Yale* bucket dredge in 1900, Frederick Powell listed the employees and their daily wages: superintendent, $5; engineer, $5; engineer and fireman, night, $3; fireman, $2–$5 two lever men, $4 each; three burlap men, $2.50 each; and one roustabout, $1.[18]

Trained as a mechanical engineer, George Burroughs's approach was hands-on. Elsewhere on the Snake, dredge owners sometimes chose a manager who had been a successful miner on a small scale, only to founder in the process of adapting to large-scale production. As one contemporary put it with his mixed metaphor: "Men have always risen on the stepping stones of the experience [of] others, but in Snake river mining they have preferred to hew their own steps as they went along until their knives dropped from their hands and they were helpless."[19]

Not so with George Burroughs. If the Burroughs-Sweetser company was very much a family affair, George was the kingpin. He had supervised the building of the *Argus*, and when it was completed, he personally kept an eye on the various experiments and improvements he and his fellow Yale entrepreneurs had devised. When the dredge began operating, he spent most of

FIGURE 3.2. The Burroughs's spacious houseboat, which moored on the shore of the Snake. Courtesy, Edgar Rice Burroughs, Inc., © 1975, all rights reserved, ™ Edgar Rice Burroughs, owned by Edgar Rice Burroughs, Inc., and used by permission.

his time on the river. Brother Harry worked closely with him, but at first he lived in Pocatello, seventy miles away. Brother Coleman also helped, but his general store in Minidoka occupied more of his time. The easygoing Ed, the youngest sibling, was a bit player on occasion.[20]

In late 1898 or early 1899, the partners constructed a substantial double-decker houseboat to provide living quarters for themselves and their families. The craft, which was baptized *El Nido* (the Nest), was sixty feet long and thirty feet wide. It contained eighteen rooms, including a thirty-foot by fifteen-foot living room and two small porches, and was moved by the same stern-wheeler steam tug that moved the suction dredge. Eventually, the houseboat accommodated up to four Burroughs brothers and their households, which included two small children, plus Lewis Sweetster and Walter Sparks and his wife.[21] On a cold December day in 1900, George Burroughs was married in the houseboat "as the waters of the tortuous Snake splashed against the side."[22] A cook and her daughter were employed to prepare meals for the family and crew, but the wives were expected to share housekeeping duties—a situation that sometimes led to so much friction and resentment that at one point George suggested the brothers split up at the first opportunity.[23] Later, in early 1903, after living facilities on a second, smaller houseboat were provided for half a

FIGURE 3.3. The Burroughs stern-wheeler tugboat moved not only the houseboat but also a coal scow up the river. Courtesy, Edgar Rice Burroughs, Inc., © 1975, all rights reserved, TM Edgar Rice Burroughs, owned by Edgar Rice Burroughs, Inc., and used by permission.

dozen or so of the dredge workers, young John Parley Byington met the girl of his dreams—the cook's daughter—while part of a dredge crew and after a lengthy courtship eventually made her his bride.[24]

In terms of logistics, with two dredges and two houseboats that required a coordinated effort on a round-the-clock basis, digging gold became a complex operation. The dredges and the tug had to be fueled and repaired; with each shift change, work crews had to be ferried to and fro; and supplies had to be provided for all. The houseboats, tug, and several lesser boats were tied up along the shore. Thus, the workforce moved with the progress of the dredges, constantly shifting its living facilities. Coleman Burroughs and his wife were delegated to search for desirable anchorage sites, close to where the family cow could be pastured.[25]

As was always the case with technology, there was much wear and tear on the gold dredge machinery and indeed on the hull itself. Grace Burroughs, Coleman's wife, recalled being awakened in the wee hours one morning by the emergency whistle on one of the dredges signaling a malfunction. All of the off-duty men immediately rowed several hundred feet to the rig to find that caulking had loosened and the dredge was taking on water. After hours of frantic labor, the leaking hull was re-caulked and dredging commenced anew.[26] On another occasion in late 1899, Coleman described a dangerous situation that occurred when he, Harry, and Walter Sparks returned from upriver with the tug heavily loaded with coal. The rudders on the vessel malfunctioned, and the wind-whipped tug drifted broadside downriver with the Burroughs brothers bailing in icy water up to their knees for several uncomfortable hours while Sparks tried to steer toward the shore.[27]

Though an exacting taskmaster, George Burroughs knew the value of satisfied, well-fed, well-housed workers; more than most employers on the river, he treated his hired hands with concern and personal attention. Crewmen serving as engineers and pilots worked eight-hour shifts, while roustabouts worked ten. Those were better hours than was common at the time, and his workers responded with efficient service. Moreover, Burroughs was an engineer with "exceptional business tact and inventive ability," whose dredging successes were "not the result of any set plan, but comprehend a good deal of careful and original thought and management on his part."[28] Each dredge engineer was expected to keep a detailed log of the dredge and to include a record of every stoppage of machinery and the reasons for that stoppage. He was also required to regularly register the readings of a set of Burroughs's automatic devices to measure the width and depth of each cut across the face of the bar and hence to calculate accurately the actual number of cubic yards of gravel handled during each shift.[29]

Both dredges were lit electrically for twenty-four hours of work and were connected with each other and with the railroad station at Minidoka by telephone, a relatively new piece of equipment. At the station, a forwarding agent with a fast team kept the company in close contact with the business sector elsewhere. As one observer put it, every aspect of the operations "has been systematized down to a fine point, and reflects great credit upon its founder. The Sweetser-Burroughs Mining Company has set the pace, and under similar competent management the Snake River Valley would seem to offer a field of great possibilities for profitable gold dredging."[30]

Realizing that good placer ground would not last forever in the Minidoka area, Sweetser and the Burroughs brothers looked elsewhere for new possibilities. George visited the Boise Basin, Warrens, Lake Creek, and the Blackfoot area—all with an eye for new technology and fresh placer possibilities.[31] As early as the autumn of 1902, he announced that he was moving his bucket dredge to the now more promising Stanley Basin, while the suction boat would be left to finish up work on the Snake.[32] With New York and Chicago capitalists as backers, George created the Stanley Basin Mining Company, of which he was president, and acquired 200 acres on Stanley Creek. Robert Green of Payette supervised the transfer and reassembly of machinery, with William Shannon as his right-hand man. When Green returned home from Robinson bar in September, early snow had made the road to Ketchum impassible, requiring a roundabout route through Challis and Mackey, while Shannon wintered in the snowbound country to continue his work.[33] By mid-April the following year the *Yale* had been completely rebuilt with the latest improved technology and was operating. A visitor to Idaho City in June 1903 noted that there were twelve feet of snow on the summit but none in Stanley Basin, adding that "the dredge of Burroughs & Co. is scooping away on their properties, that being the only thing at present in operation there."[34] The operation was short-lived; for whatever reason, it was not a success and soon closed down.

Also onto the scene in 1903 came Ed, the youngest of the Burroughs brothers, twenty-eight years of age. No stranger to southern Idaho, as a frail teenager Ed had been sent by his parents during a flu epidemic in 1891 to spend time on the Sweetser-Burroughs ranch, an experience he enjoyed. He returned briefly to help on the dredge and with the cattle in the spring of 1898. But for the most part, young Ed was something of a misfit. After one semester, he dropped out of Phillips Academy because of poor grades and subsequently went to a military school in Michigan, where he also later taught for a few years. Appointed to West Point in 1895, he failed the entrance examination. At one point he joined the Seventh Cavalry but found army life in Arizona not to his liking and was discharged, thanks to parental influence. He tried to run a book and stationary store in Pocatello but soon gave it up and was back working in his father's automobile battery business in Chicago when he joined his brothers' work crew in the Stanley Basin.[35]

It is doubtful that the lackadaisical Ed added much to the labor force. Evidence indicates a hesitancy on his part to pull his weight where the dredge was concerned, and his romantic walks along the stream in the scenic Sawtooths apparently brought words from brother George about his flimsy work ethic. The result was that Ed, his wife, and their collie soon left Stanley

Basin for the western Snake, near Parma, where Harry Burroughs was running another boat.[36]

In mid-1902 George had indicated that his company would put a dredge on the river near the Idaho-Oregon border and that men and teams were already at work hauling machinery.[37] It is not clear whether this was a new boat or a modification of an existing one, but it was not the *Yale*. In late 1903 J. T. House had been operating a dredge for the Moraine Mining Company on the Snake thirteen miles from Ontario on the Oregon side. He reported that the company had been getting from seventeen cents to thirty-five cents a yard from the gravel, an output considered excellent for a river dredge.[38] Soon, Harry Burroughs emerged as superintendent of the Moraine Mining Company, with a boat working upstream from Ontario in the final months of 1903. Harry had attended a Chicago meeting of the board of trustees at the end of the run in the winter and publicly expressed his confidence that the company's "properties in the Big Bend will develop into a bonanza."[39] As did many of the crewmen, the newly arrived Ed Burroughs and his wife lived on the Idaho side of the river in Parma, and Ed seemed to have applied himself to his work more than he had in the Stanley Basin—or perhaps Harry was not the strict taskmaster that George was. When Ed was elected a trustee of the Village of Parma in April 1904, it was a short-lived victory. Soon thereafter the Moraine dredge closed for good.[40] That left only the original suction boat farther east in the Minidoka area, and according to one contemporary observer, operations there were suspended two years later.[41]

One version of the collapse was that the Burroughs brothers went bankrupt because of their venture into the Stanley Basin—that an eastern syndicate had persuaded them to take over placer property there and Sweetster-Burroughs had gone under after being unable to meet its obligations. Another explanation is that the company had bought new ground somewhere on the Snake and went through an arduous and lengthy move of its equipment through dangerous rapids, only to find that the property had been salted.[42] Given the lack of evidence to support such hearsay, the reason for the demise was likely much simpler: the ground the company was working was no longer profitable.

In any event, the Sweetser-Burroughs gold-dredging saga was finished, and the main characters scattered. Lewis Sweetser stayed in Idaho, where after two terms in the legislature he was twice elected the state's lieutenant governor and might realistically have become governor had it not been for the fact that his wife was rumored to be an alcoholic.[43] George, Harry, and Coleman Burroughs closed out their Idaho holdings and went their separate ways, in general ending up back in Illinois. So did young Ed. For the next six to eight years he tried a variety of jobs ranging from railroad policeman, accountant, salesman of lightbulbs to janitors, hawker of pencil sharpeners, an Alcola (a cure-all for alcoholism), and selling the popular *Stoddard Lectures* door to door. Eventually, he found his true calling. Beginning in 1912, writing under his full name, Edgar Rice Burroughs, he published twenty-four Tarzan novels plus fifty more books, most of them early science fiction set on Venus, Mars, and the center of the Earth. He sold an estimated 30 million to 60 million copies in thirty languages and was considered by experts to have been "the most widely read American author of the first half of the twentieth century."[44]

The waves of success enjoyed by Sweetser and the Burroughs brothers radiated widely and no doubt encouraged others in their attempts to dredge the Snake. In his annual reports and in mining periodicals, Idaho inspector of mines Robert N. Bell was a hearty and vocal champion during their early years, giving detailed descriptions of both the suction dredge and the bucket-line boat that circulated through mining periodicals and local western newspapers and reached professional engineers as far away as Newcastle-upon-Tyne in England.[45] But Bell was careful to point out that Sweetser-Burroughs was the exception on the Snake and that while the company had enjoyed prolonged success with the suction boat, its bucket-line dredge had not been profitable. Moreover, he was aware that the river bars were inconsistent in content and erratic in production—and there were always problems with recovering and processing the fines and the black sands. In 1907, after Burroughs and Sweetser had shut down, Bell noted that there had been "probably $1,000,000 worth of dredging plants built along the Snake, but this is the only one that ever paid a dividend."[46] To be sure, the *Argus* had paid one $10,000 dividend—which the pubic was not allowed to forget—but the *Yale* bucket dredge was never profitable.[47]

George Burroughs was willing to share information about his dredges with the press—to a point. As he pointed out in the last paragraph of a long descriptive article on the *Argus* in the spring of 1897, he did not show everybody the boat. He did not choose to give away valuable knowledge gained by trial and error and experience. For example, there were tricky questions such as how he was able to reinforce the pump so it would last a year instead of a week or what was the exact pitch required of the burlap tables for maximum efficiency.[48] At times, Lewis Sweetser could be equally coy about the machinery. In 1901, while discussing the company's success in saving flour gold, a matter of considerable speculation among mining men, he informed a Boise reporter that "the process was largely a secret of his firm's, the result of several years of expensive experiments."[49]

Sweetser and George Burroughs especially were newsworthy public figures. Sweetser, after all, was a legislator and, later, lieutenant governor. Burroughs was often cited in the press as the model of what a mining man should be. Governor Frank Steunenberg appointed him to be one of the state's official delegates to the third annual session of the International Mining Congress in Milwaukee in the spring of 1900.[50] The Boise *Statesman* carefully tracked the progress of the company's dredging operations, both through items borrowed from other newspapers and through brief interviews when either man was passing through the state capital. The typical interview usually included a positive phrase such as "he reported [that] business [is] good with his company which has been securing very satisfactory returns."[51] Even Salt Lake City editors seemed to have a lively interest in Snake River dredging activity. In the summer of 1897 the *Tribune* described a three-day run in which the Sweetser-Burroughs *Argus* produced "a slug of gold"—thirty-eight troy ounces, worth about $760, from sands averaging 5 cents to 8 cents a yard.[52] When George Burroughs visited the Blackfoot placer fields five years later, he was lauded by the *Salt Lake Herald* as "a gold dredging engineer of exceptional ability," whose advent into that area "is hailed with anticipation."[53] Burroughs did not invest there, however. Instead, he went to the Stanley Basin.

There is no doubt that the Sweetser-Burroughs operations furthered the hopes of others. They provided a prototype for success. What they could do, others could emulate, especially with newer and better equipment to revolutionize the industry. On the Snake, success never happened. Dredge expert J. P. Hutchins commented in 1906 that it "is a difficult matter *not* to save a large percentage of the gold in most placers—almost impossible to lose it, in other words—while, on the other hand, it is almost impossible to save the gold in some deposits, like those of the Snake river, Idaho, when operation is conducted on a large scale."[54]

Notes

1. See John Taliaferro, *Tarzan Forever: The Life of Edgar Rice Burroughs, Creator of Tarzan* (New York: Scribner, 2002), 49–50; Irwin Porges, *Edgar Rice Burroughs: The Man Who Created Tarzan* (Provo, UT: Brigham Young University Press, 1975), 79; Phillip Burger, "Sweetser and the Burroughs Boys," *Burroughs Bulletin* 19 (New Series) (July 1994): 14–15.

2. *Engineering and Mining Journal* 63 (June 12, 1897): 610; *Engineering and Mining Journal* 73 (February 15, 1902): 341.

3. The Albion *Busy Bee*, quoted in the *Idaho Daily Statesman* (Boise), May 5, 1895.

4. *Engineering and Mining Journal* 60 (October 12, 1895): 353. The *Engineering and Mining Journal* estimated that the dredge saved about half of the gold content raised from the low bars and along the edges of the stream. *Engineering and Mining Journal* 73 (February 15, 1902): 341.

5. *Idaho Daily Statesman* (Boise), May 5, 1895.

6. *Engineering and Mining Journal* 70 (October 6, 1900): 395; *Transactions of the American Institute of Mining Engineers* 18, May 1889–February 1890 (New York: American Institute of Mining Engineers, 1890), 603.

7. *Engineering and Mining Journal* 63 (June 12, 1897): 610; *Engineering and Mining Journal* 73 (February 15, 1902): 241.

8. Ibid. 63 (June 12, 1897): 610; Robert Bell, *Report of the Mining Districts of Idaho for the Year 1904* (Boise: Statesman, n.d.), 47.

9. *Engineering and Mining Journal* 73 (February 15, 1902): 241.

10. *Idaho Daily Statesman* (Boise), March 4, 1898.

11. *Engineering and Mining Journal* 70 (October 6, 1900), 395–96; *Engineering and Mining Journal* 73 (February 15, 1902): 242.

12. Ibid. 70 (October 6, 1902): 395–96.

13. *Idaho Daily Statesman* (Boise), August 25, 1900.

14. Ibid., November 16, 1898.

15. Ibid., December 11, 1899.

16. *Engineering and Mining Journal* 73 (February 15, 1902): 242.

17. Ibid., 241.

18. Ibid. 70 (October 6, 1900): 396.

19. *Mining and Scientific Press* 80 (February 24, 1900): 206.

20. Taliaferro, *Tarzan Forever*, 50.

21. Ibid.; Porges, *Edgar Rice Burroughs*, 81.

22. *Idaho Daily Statesman* (Boise), January 12, 1900.

23. Porges, *Edgar Rice Burroughs*, 711.

24. Ibid., 81; "John Parley Byington," http://www.auntromas.com/john_parley_byington.htm.

25. Taliaferro, *Tarzan Forever*, 50.

26. Porges, *Edgar Rice Burroughs*, 82.

27. Coleman Burroughs to Grace Moss, December 1, 1899, quoted in ibid., 710–11.

28. *Engineering and Mining Journal* 73 (February 15, 1902): 242.

29. Ibid.

30. Ibid.

31. *Idaho Daily Statesman* (Boise), November 16, 1898; November 8, 1901; *Salt Lake Herald*, March 17, 1902.

32. *Idaho Daily Statesman* (Boise), September 17, 1902.

33. Ibid., November 30, 1902; *Salt Lake Mining Review* 5 (May 25, 1903): 13.

34. *Idaho Daily Statesman* (Boise), April 12 and June 19, 1903.

35. George T. McWhorter, "Burroughs, Edgar Rice," at http://www.anb.org/articles/16/16-002230.html, *American National Biography Online*, February 2000; Taliaferro, *Tarzan Forever*, 14, 58–60.

36. Porges, *Edgar Rice Burroughs*, 86.

37. *Idaho Daily Statesman* (Boise), June 10, 1902.

38. Ibid., November 28, 1903; "Mineral Resources and Mineral Industry of Oregon for 1903," *University of Oregon (Eugene) Bulletin* 1 (New Series) (1904): 21.

39. "Mineral Resources and Mineral Industry of Oregon for 1903," *University of Oregon (Eugene) Bulletin 1* (New Series) (1904): 21; *Idaho Daily Statesman* (Boise), November 28 and December 3, 1903; *Engineering and Mining Journal* 77 (May 5, 1904): 728.

40. Porges, *Edgar Rice Burroughs*, 88–90.

41. Henry G. Granger, "Hydraulic Dredging for Gold-Gravels," *Bulletin of the American Institute of Mining Engineers* 28 (April 1909): 349.

42. Porges, *Edgar Rice Burroughs*, 90.

43. *Idaho Daily Statesman* (Boise), December 29, 1908; September 27, 1910. An excellent summary of Sweetser's post-dredging career in Idaho and Montana is in Burger, "Sweetser and the Burroughs Boys," 18–22.

44. McWhorter, "Burroughs, Edgar Rice"; Taliaferro, *Tarzan Forever*, 14, 58–61.

45. Robert Bell, "Dredging for Gold in Idaho," *Engineering and Mining Journal* 73 (February 15, 1902) 241; Bell, *Report of the Mining Districts of Idaho for the Year 1904*, 47–52; *Mining and Scientific Press* 94 (April 27, 1907): 542; North of England Institute of Mining and Mechanical Engineers, *Transactions* (Newcastle upon Tyne: Andrew Reid, 1907) 54:12–13, vol. 26 (1903–4); *Transactions of the Institution of Mining Engineers* (Newcastle upon Tyne: The Incitation, 1905) 26:616–17.

46. Quoted in *Mining and Scientific Press* 94 (April 27, 1907): 542.

47. Originally, Bell stated that the suction dredge had paid two $10,000 dividends, but he subsequently changed it to only one. Bell, *Report of the Mining Districts of Idaho for the Year 1904*, 47, 52; *Mining and Scientific Press* 94 (April 27, 1907): 542.

48. *Idaho Daily Statesman* (Boise), April 20, 1897. This same article appeared verbatim in the *Engineering and Mining Journal* 63 (June 12, 1897): 610, but without this final cautionary paragraph.

49. *Idaho Daily Statesman* (Boise), January 11, 1901.

50. Ibid., February 4, 1900.

51. See, for examples, ibid., November 16, 1898; December 11, 1899; January 11 and December 8, 1901; August 23, 1902.

52. *Salt Lake Tribune*, August 2, 1897.

53. *Salt Lake Herald*, March 17, 1904.

54. John Power Hutchins, "Gold Dredging in 1906," in *The Mineral Industry: Its Statistics, Technology and Trade during 1906*, edited by John Power Hutchins (New York: Hill, 1907), 15:428.

4

Pierce

Idaho's golden age was in the 1860s, when restless prospectors discovered many of the placer deposits and quartz veins in the western and central parts of the territory. Historian Hubert Howe Bancroft made the statement that "the miners of Idaho were like quicksilver. A mass of them dropped in any locality broke up into individual globules, and ran off after any atom of gold in their vicinity. They stayed nowhere longer than the gold attracted them." In 1860 Elias Davidson Pierce, a trader among the Nez Perce, first found gold on one of the tributaries of the North Fork of the Clearwater River—on tribal lands. The result was an influx of gold seekers from Walla Walla and the establishment of Pierce City in December; a few months later and just two miles away, Orofino was founded at a time when there were 7,500 men in the area. What became the town of Lewiston at the juncture of the Snake and the Clearwater developed as a jumping-off place for the new mines. In the spring of 1861 a small band of prospectors encroached further south into Nez Perce territory and found shallow but very rich placers on the upper South Fork of the Clearwater, 125 miles from Pierce City, at a site they named

DOI: 10.5876/9781607324751.c004

Elk City. Another small group left Orofino in July of the same year and found incredibly rich diggings a few miles north of the Salmon River—in modern terms, about fourteen air miles east-northeast of present-day Riggins. Despite inclement weather and almost impossible travel conditions that continued into the winter, thousands flocked to the new settlement of Florence, as it was called.

Not far away, James Warren's party of prospectors also found paydirt, and in 1862 Warrens soon became a bustling camp. One more important mining discovery was an outcome of the constant search in western and central Idaho—the riches of Boise Basin, thirty miles northeast of the capital city. Moses Splawn had been in Florence and Elk City earlier and in 1862 headed the little band of miners who found high placer values. By that autumn half a dozen towns had sprouted, including Bannack, Placerville, Centerville, and Hogem; by 1863 the rush was on, with a peak population in the mid-1860s estimated at between 25,000 and 40,000.[1]

West-central Idaho won its reputation as one the country's great mining districts primarily through production from these early camps. Pierce City, Elk City, Florence, and Warrens were all located in hard-to-reach high placers with faulted valleys, and they all followed a similar pattern of development. Their placer gold tended to be large, rich particles in shallow gravel and easily obtained. They played out in a few years, and the miners moved on to the next camp. In all cases they were replaced by Chinese who worked the ground well into the 1880s and 1890s. In all instances attempts were ultimately made to apply dredge technology, sometimes with no success and sometimes profitably. The Boise Basin provided more gold and showed more stability over a longer period of time than other regions, in part because the warmer climate permitted an easier family life.

The Pierce district was the first of the high placer deposits. As was the case in many placer areas, talk of installing a dredge or dredges often began in the late 1890s, not long after the early successes in Montana and California. First came the dreamers. So it was with the Pierce City and Orofino districts, both at an altitude of not quite 4,000 feet. In the spring of 1898 Jack Dunn, a visitor from Wallace, was quoted as saying that Orofino was destined to become "one of the greatest districts in Idaho" because of a "new departure"—"the deep placer dredger," which had gone beyond the experimental stage. Thousands of acres along the river bottom were available and were so flat that they were impossible to work by sluicing. Some of the ground had proven to be rich, but it was so deep to bedrock that it discouraged hand labor. Moreover, the bedrock was soft and easily dug into by a powerful bucket line. Cheap electric power could readily be harnessed from Orofino Creek, where it entered a narrow canyon before plunging into the Clearwater. Already, said Dunn, representatives of eastern capitalists had shown an interest, and a number of locations had been made for potential dredging sites. Dredge experts had seen the opportunity for ten to twelve boats within a few years and predicted that the new technology would allow the district to surpass all others, even the Boise Basin.[2] It did not happen the way Dunn hoped.

In 1903 there was positive action in the Pierce City district when a firm calling itself the Idaho Dredging Company posted its first dividend of $10,000.[3] At the time, the Pierce district, a separate entity, was located in the southern part of Shoshone County. Beginning in 1904, that county was reduced in size, and parts of the rich placers—including Orofino Creek, Orofino, and Pierce City—were attached to Nez Perce County. In 1911 the region became part of Clearwater County.[4] Since there is evidence that Shoshone County had three or four Risdon dredges and at least one other by an unknown manufacturer by 1903, it is difficult to ascertain whether the Idaho Dredging Company represented a new machine or one shifted from farther north in the Coeur d'Alenes.[5] What kind of machine, who owned it, and precisely where and how long it operated are questions yet unanswered. Even the name of the company is not clear.

Subsequently, after a few months had passed, a spokesman for a new, yet-unorganized corporation gave promise of additional dredge boats. Wellington S. Harris, a British Columbia mining investor who was also interested in placer ground near Pierce City along Orofino Creek, spoke of a company to be formed during the winter and machinery to be installed the next spring, 1904. Six prospecting shafts had been dropped to bedrock, Harris reported, and tests showed values of up to twenty cents a pan to a phenomenal seven dollars per cubic yard. The proposed new company planned to build four large electric dredges to dig both virgin gravel and some worked years earlier.[6] Like many other ambitious proposals before and after, Harris's dreams never materialized. He was later involved in several successful Canadian mining enterprises but not in the Pierce City district.

Federal documents in 1905 commented that since 1904, one of the two dredges digging in the Delta district in the Coeur d'Alenes had moved to near Pierce in Nez Perce County and would be ready for operation the following spring.[7] This was one of the several three-foot Risdons in Shoshone County between 1900 and 1902, no doubt the one owned by the Delta Gold Mining Company.[8]

The Idaho Company, Ltd., was organized in 1906 with nominal capital of $100,000, mostly from Chicago, and an all-embracing charter that included not only the right to pursue mining and dredging but also a long litany of other business activities, ranging from smelting, lumbering, and generating electricity to general manufacturing and merchandising. Its president was Marquis Eaton, an eminent Chicago lawyer, and the resident agent in Pierce City was H. Galligan, previously in charge of the Delta boat on Trail Creek in the Coeur d'Alenes.[9] For all of the company's many possibilities, gold dredging stands out as its most successful and enduring commercial enterprise.

The boat was moved by rail from the Coeur d'Alenes to Greer, then to the rail head a few miles south of Orofino, then it was hauled to Pierce by wagon in January 1906. It began digging in mid-May of the same year on 450 acres at and near the confluence of Orofino, Rhodes, and Canal Creeks. More than half of that land was deemed dredge-able and was said to have paid a dividend of $10,000 in its first year. This was the high placer plateau gravel so common around Pierce City. Roughly 80 percent of the workable ground consisted of tailings from earlier placer operations, and recovery was said to have run close to the 21 cents per yard shown in extensive testing. Operating costs were reported to be 8 cents a yard.[10]

After a very successful 1907 season and having acquired additional ground, the Idaho Company, Ltd., was contemplating adding a new dredge. Accordingly, it sent well-known dredge engineer John Power Hutchins to Pierce to test the drill ahead of the old Risdon and to give a general assessment of its work. Hutchins lauded Paul Hare, who was in charge of the machine, and his crew for doing good work with a woefully inefficient rig. The dredge was of faulty design and made of poor materials, he said. New, improved close-connected buckets would save 25 percent of the cost of fuel and repairs. Hutchins urged building a machine shop, adding quicksilver riffles to at least part of the gold-saving tables, and ventilating the interior of the hull to lengthen its life. He also recommended a number of ways to lessen wear and tear on the tumblers and buckets and to prevent spillage, faulty dumping, and poor washing. The present dredge was meant to dig deeper than the relatively shallow thirteen or fourteen feet to bedrock at Pierce, and it lost much gravel in the digging. In California, dredges of this type had been scrapped. By all means, advised Hutchins, get a new dredge, but get a modern one. The report also included results of the test drilling.[11] Seven years later, the operator of another boat on the same ground made this comment: "Needless to say, there is a very wide discrepancy between the anticipated recovery as shown by this report and the actual [recovery]. In fact, Mr. Hutchins must have been a bridge engineer and slipped in a factor of about four—that is a factor of four times as good so they would be sure and get enthused."[12]

The second dredge of the Idaho Company, Ltd., completed in November 1909, was built by New York Engineering, a firm headed by Albert C. Ludlum, which had a reputation for building excellent machinery. With four-foot close-connected buckets, it was capable of handling 2,000 yards per day. Its cost was estimated at $50,000 to $60,000. It was reported that the new steam-powered dredge was equipped with "five engines, so adjusted that a breakage in one part of the machinery would not necessarily close down the work of the entire boat." That was probably a misinterpretation. Very likely, this was a reference to different gear mechanisms, each designed to power a different component, such as the bucket line, grizzly, gold tables, or winches for stepping forward. It is obvious that if any single vital part of that sort were not functioning, the entire dredge would have to be shut down. Superintendent Hare believed the boat would be able to run through the winter months and that the two dredges would last nine more years.[13]

The new machine was set to work on ground adjoining Pierce City—practically downtown—in early December. The property had been acquired four years earlier, and the Risdon had started up on it in 1907, with considerable success. At the end of the 1909 season, the three-foot Risdon had just completed its fourth season of digging, having worked ten months in each of those years. While the company would give no official figures on production, it was well-known that regular and substantial shipments of gold had been sent to Boise. Idaho state inspector of mines F. Cushing Moore estimated that the smaller machine had produced about $45,000 in 1909 alone.[14] By the beginning of 1911, the older dredge had been leased to the Sprague Mining and Milling Company and operated somewhere in the same area for three years until it ran out of ground in 1913.[15]

For some reason, the new boat of the Idaho Company, Ltd., was idle during most of 1912. Moore believed poor financial decisions by absentee eastern capitalists were responsible. Located on excellent ground and estimated to average thirty cents a yard, the dredge was poorly designed, with a steam operating plant so expensive that it left little room for profits. With electric power now available in the district, Moore urged the firm to spend a few thousand dollars to replace the old steam plant with a modern electric system. It would take a change of ownership to accomplish this, however. With $6,000–$7,000 in liens for unpaid labor against it, the company fell on difficult days, and Rupert H. Bailey, one of the shareholders, bought the boat at a sheriff's sale in Orofino sometime in 1913. While being overhauled in September of that year, it caught fire and was destroyed. At that time the company was known as the Pierce Company, Ltd., and the owners were listed as Rupert H. Bailey of Walla Walla, N. H. Hare of Spokane, and Paul Hare of Pierce, who promptly announced that the machine would be reconstructed.[16]

And resurrected it was: the boat was electrified, and it started up in March 1914, but when World War I erupted in Europe later that summer, the company lost a dozen or so workers. For whatever reason, it operated for only a truncated season in 1915 as the Idaho Power Company dredge, under Bailey's management. Thereafter, under the name Pierce Company, Ltd., dug full seasons every year until 1920, except for 1918 when it was idle. The boat worked at least two more years as the Bailey dredge on Orofino Creek, much of the time under superintendent H. J. Waters.[17] When he was proposed for membership in the American Institute of Mining Engineers in 1916, Rupert Bailey, a graduate of the University of Nebraska and head of a construction firm in Walla Walla, was listed not only as the owner and operator of Pierce Company, Ltd., but also as the manager of the Idaho Placer Mining Company, a large hydraulic mining outfit at Newsome, a few miles northwest of Elk City.[18]

Later, Bailey liked to reminisce about what was unearthed besides gold. Going over ground previously worked by hand by, first, white miners and then by Chinese, the dredge "dug up just millions of Chinese coins," not to mention a few Chinese razors, he said. "Of course we dug up all the bullets ever shot in the last fifty to one hundred years or since the old timers came. It made it difficult to clean up, as we had so much of this lead."[19]

FIGURE 4.1. Rupert H. Bailey was a successful owner and operator in the Pierce area for eight years. Courtesy, Idaho State Historical Society, Boise, ISHS 72-112.b.

In 1915 eight of the boats operating in Idaho, four of them in the Boise Basin, closed down for various reasons while the Pierce Company dredge continued as the main single producer of gold in its district, with limited competition. Wellington Harris was back in 1914 and promised to have a $75,000 boat running twenty miles west of Orofino by August 1, 1915. Perhaps he did so—briefly. The American Placer Mining Company of Spokane, with J. M. Porter at the helm, started up its three-foot Risdon in 1915. Porter was also president of the Spokane firm, which was nominally capitalized at $500,000 and had 230 patented and 90 unpatented acres along Orofino Creek. After a brief test run, he decided that some serious changes were required before the Risdon could continue. Instead of the dredge, the company mined with a hydraulic lift and sluice in 1916 and 1917.[20] In 1920 Porter was both president and manager of a new concern—the Gold Dollar Mining Company—and in charge of running the irrepressible three-foot Risdon on Orofino Creek.[21]

By the beginning of 1918, the Pierce Company had concluded its digging at Pierce City and was moving its base several miles up the creek, where spokesmen believed there was ground that would last five years and that had twice the gold content. The power plant was five miles farther up the stream. As in Alaska, snowfall and low temperatures facilitated transport of lumber and machinery.[22]

Five placers in Clearwater County produced $7,385 in gold and 89 ounces of silver in 1920, a decrease of 37 percent from the previous year. All the gold came from the Pierce district, and most of it came from dredging. A newcomer on the scene, the New Era Gold Dredging Company, with Edward Gaffney as president and John Gaffney as secretary, owned 320 acres of placer ground in the district but seems never to have been operational. Its property was under lease and bond to Rupert Bailey. The take the next year was less than half of the previous year's take, and most of it came from the Bailey boat, which was also responsible for much of the gold, valued at $5,882 in 1922.[23]

At the time, Rupert Bailey still retained his mining interests and resided in Pierce—indeed, he represented Clearwater County in the state legislature that year—but in 1922 James W. Hall of Winslow, Washington, had replaced him as owner of the Bailey dredge. Within a few years he had moved to Seattle. Three men from that city initially bought his dredge but for some reason were unable to continue to run it. The venerable Bailey dredge, as the

New York Engineering Company machine continued to be called, was again rebuilt in 1922 and a new hull was constructed. It was then moved to Rhodes Creek, where active digging began in September, but it ran only briefly in 1923. Backed by Seattle and Spokane capital, the Bole Mining Company was said to be interested and had already begun work on a group of claims along the stream.[24]

The output of placer gold had fallen off, and in 1926 the US Bureau of Mines lamented the fact that from 1911 to 1917 the district had been an important producer of the precious metal, primarily because of dredging. The bureau hoped renewed interest would be awakened by construction work being done by the Northern Pacific Railroad on the branch line from Orofino to Pierce. Two years later, after the rail line had been completed, the total placer gold yield was valued at only $266.[25]

In mid-1929 J. M. Porter, the Spokane engineer, and two co-owners of placer claims in the Pierce district, John Gaffney and E. E. Steele, made preparations to dredge the ground. The property was on Rhodes Creek, and the dredge was owned by Mat Erb of Briar and now on adjacent land leased to Erb. The two properties were to dredge jointly, with a hoped-for trial run in the autumn. L. F. Sperry and associates of Portland were also interested in the project.[26] Unfortunately for the participants, the plans fell through, probably from lack of funding after the stock market crash in October of that year.

In July 1932 another attempt was contemplated by a different group. H. H. Smith, who, like Porter, was a Spokane mining man, announced in Pierce that the New York–Idaho Dredge Corporation was converting the old Bailey dredge to diesel power and hoped to have it in operation within thirty days. Smith did not indicate a specific location but dwelled on the richness of the ground, which averaged forty cents per cubic yard, and the boat, which could handle three yards a minute with a crew of eight to ten men. A. V. Arneiter was the manager. The machine was one of four bucket dredges working in Idaho in 1932.[27]

It took veterans of the Klondike and Alaskan mining and dredging scene to achieve some success in tapping the riches of Rhodes Creek. Norman C. Stines was an experienced engineer who cut his teeth on Russian dredging and went on to play a significant role in the formation of the giant United States Smelting Refining and Mining Company, which came to dominate the Alaskan dredge field. Now, in late 1932 Stines organized Gold Dredging, Inc., to take over the New York–Idaho Dredge Corporation property near Pierce on Rhodes Creek, which had been revived by A. V. Arnreiter. Associated with Stines were Sam Godfrey and Ben Bromberg, also seasoned Alaskan miners.[28]

Godfrey stirred up excitement in Pierce. After his experiences in the far north, he had wandered around a number of the old Idaho mining districts from which millions had been taken out in the early years and that were now considered worked out. He came to the conclusion that on Rhodes Creek the old-timers had not reached bedrock, where the richest values were located, and that a dredge was the answer to recover the remaining treasure. Short on funds, Godfrey interested a few private investors, like Leon and Dorothy Sabo; he then obtained the old Bailey boat and moved it piece by piece to a site about a mile from town. There, after overcoming a number of obstacles, it was rebuilt in a rather crude way and, after fits and starts, began operations. It was an immediate success. Sam Godfrey flashed gold nuggets around Pierce City and made much of the fact that the machine had cleaned up $22,000 in precious metal in seventy days. The company filed on 280 acres of additional claims and acquired mineral rights to about a half a mile more along the stream.[29]

Apparently, the dredge was too fragile for sustained digging and again had to be completely reconstructed and modernized before the 1934 season. Godfrey had a new partner in the enterprise, Herb Klough of Everett, Washington. Some of the machinery still came from the antique Bailey dredge, but the larger part, including the hull and a 165 horsepower Western Enterprise marine-type diesel engine, was new. The diesel not only powered the boat and all of its equipment, it also generated electricity for lighting, welding, and other uses. A heavy steel-pointed spud replaced the old-fashioned guy cables to hold the dredge in place. New winches were added to control various pieces of equipment, along with a high-pressure centrifugal pump for the grizzly and a save-all to retain about 10 percent of the gold lost when the buckets emptied into the hopper. The bucket line was completely overhauled, although the venerable upper tumbler was retained. It was a period piece; it was estimated to have been in the district for thirty years and to have been used on seven other dredges. Dredge master Joe Linkey had charge of the rebuilding, which involved a crew of twenty men. Gordon Jones of Pierce, head mechanic for the job, worked forty-five days on electrical welding and used more than sixty drums of oxygen and over half a ton of welding rods in the process. The completed boat cost about $70,000, and its owners anticipated a six-year run.[30]

The *Virginia Ann*, as the boat was called, began digging in August 1934 under Linkey's guidance, but he soon moved to Alaska to take charge of another dredge and was replaced by William Highland, said to have had dredging experience "virtually all over the world." Operations seemed generally profitable, although at one point excessive black sands hampered recovery. For a month or so in the spring of 1935 it was hard going when the machine had to cross a strip of solid rock, and dynamite was required to blast a channel ahead of it. At this time, Sam Godfrey still owned a controlling interest in the company.[31] A photograph taken later in the fall of the same year shows the boat "scooping up mud and sand" from the creek bed, handling 2,000 yards a day on the ground, and averaging twenty cents a yard, or $400 a day. A subtitle to the newspaper story stated that the "*Virginia Ann* is a gold digger."[32]

There was another entry in 1935, when Patterson Mines, Inc., announced that it had acquired dredging rights on 2,000 acres on Quartz Creek four miles north of Pierce. Testing on the ground had averaged eighty cents a cubic yard, said an official report. The company planned to operate "with a Link belt gas shovel coupled by a 12-foot steel pole to a gold-saving machine weighing 15 tons mounted on caterpillar tracks, making the entire plant mobile." Complete with a hopper and a trommel screen 28 feet long and 5 feet high, the equipment was fitted with four sluice boxes with Hungarian riffles built under the screen and a 50-foot stacker to carry the tailings. When assembled, the machine was about half a city block long. All this was powered by a 45 horsepower engine. Patterson Mines' publicity man closed his brief newspaper article with a familiar sentence: "It will take many years to complete the acreage."[33] There was no more publicity from the firm.

The Pierce district was the only important producing district in Clearwater County in both 1935 and 1936, mainly because of the gold-dredging boat, although its output in 1936 was about 500 ounces less than it had been the previous year. The machine was still digging in 1937, as a news story indicated, producing not only gold but an interesting supply of antique bullets, scrap iron, and square-cut nails. The *Minerals Yearbook* pointed out that a new bucket dredge, owned by the Gold Creek Placer Company, also started up on Orofino Creek in the district in 1936 but was too late in the year to have had much impact.[34] The editor was mistaken about the bucket dredge. The machine was a dragline with a floating washing plant, a new type of mining equipment that was cheaper than traditional dredges and was able to work in more restricted areas. This more modern technology was quickly applied throughout the placer mining industry, sometimes with success and sometimes not.

Still in its formative stage, Gold Creek Placer had acquired 230 acres of the Orofino Creek property of the late Spokane engineer J. M. Porter. Before his death, Porter had transferred the ground to his American Placer Mining Company, which in turn sold it to Gold Creek Placer on a 15 percent royalty contract. Corporate officers were headed by president Elmer E. Johnston, an eminent mining lawyer and member of the Washington State Legislature. Amid adoring fanfare from the local press, much was made of a new experimental "gold machine" being tried on the ground—apparently a new approach to separate the precious metal. Planned by two mechanical engineers in Coeur d'Alene and approved by one of the leading placer mining experts of the day, H. G. Peake of San Francisco, the new equipment was to be constructed at Union Iron Works in Spokane. Soon, "virtually every department" of the firm and "a majority of the employees" were occupied with the project. Description of the machinery is vague, save that it would be seated on two steel pontoons, each twenty-eight feet by thirty-six feet and weighing a total of twenty tons. The "gold machine" would be powered by three Fairbanks-Morse diesel engines, 60, 40, and 20 horsepower, respectively, at a total cost of about $7,000. The digging would be done by a new 2-cubic yard, 32-ton Lima shovel, with a 60-foot beam and a 165 horsepower diesel power plant, to replace the three-quarter-yard shovel currently on the property. The ground itself stretched along Orofino Creek for several miles and was located about fifteen miles east of Orofino.[35]

Like most operations of its kind, Gold Creek Placer provided floodlights for round-the-clock work. Unlike most of its counterparts, it sought to provide incentive pay to some of its workers. In a two-shift arrangement, the firm employed a superintendent, two shovel operators, and two gold machine men—all of whom received $5.00 a day, plus board and lodging in a camp that accommodated fourteen persons. If they handled 1,000 yards a day, each worker would receive an additional $2.50 a day; if 2,000 yards, $5.00 additional; if 3,000 yards, $12.50 more; and if 4,000 yards, $15.00 a day extra. Spokesmen for the company believed that with the new equipment, each shift would be able to manage 2,500 yards.[36]

How effective this bonus may have been is not documented. The new machinery worked only briefly in 1937, from August 18 to November 20, during which period it extracted $14,493 worth of gold, according to the firm's annual minutes. At this time, the company doubled its capitalization to $100,000. In 1938 the machinery again ran only a few months but was the largest gold producer in the Pierce district. Thereafter, there was no mention of the Gold Creek Placer Company in the *Minerals Yearbook*.[37]

Arthur Campbell, Idaho's state mine inspector, lauded the achievements of Gold Dredging, Inc., in July 1936 and pointed out that on the old Pierce flats there were "millions of yards of gravel" for similar exploitation. Worked over several times, other favorable placer properties were also available, information on which could be obtained from Judge Steele of the probate court in Orofino.[38]

In 1939 the Dividend Placer Mining Company put a one-and-a-half-yard dragline and floating washer on the American property but suspended operations after a short run. The overall gold output of the Pierce district was more than double that of the previous year, thanks to the operation of a new two-and-a-half-foot bucket-line all-steel dredge owned by the Quartz Creek Dredging Company. The president of the firm and designer of the new Quartz Creek boat was Walter W. Johnson, well-known for building such machinery in the West and in Alaska. The company was incorporated in March 1937, with capital of $150,000, and it immediately acquired rights to the ground from William Harris, proprietor of the Westminster Hotel in Spokane, who had long been interested in mining in the Pierce area. The dredge worked in the stream from which the firm took its name from July 8 to the end of the year and handled 495,639 yards of gravel. By September 1940 it reported that the boat was continuing its work up Quartz Creek and then up Trail Creek. The dredge operator reported that he recovered 94 percent of the values, with his results fully reflecting his test data. But alas, placer gold production dropped 45 percent in 1941 because the Quartz Creek Dredging Company suspended dredging in March. Only the operation of two draglines prevented a larger decrease.[39]

One of the draglines originally belonged to Jett Ross Mines, Ltd., previously the second-largest producer of gold in Wyoming, operating in Big Atlantic Gulch in Fremont County. This rig had a 1¼-cubic-yard bucket and a 50-foot boom, with a washing plant made of wood and set on skids. The owner,

Charles Ross, had spent his adult life in placer mining. In mid-1937 he signed a deal with William Harris for rights to land on Quartz Creek, in which he agreed to pay Harris $90,000 in royalties as the money was taken out in gold. On Quartz Creek the dragline worked briefly but was idle in 1938. The following year it was leased to another firm in which Jett Ross Mines had a 20 percent interest and continued to dig until March 25, 1941, when it ran out of ground. Since the machine needed repairs and parts were almost impossible to obtain during the war years, it was put into storage. The company that leased it was unidentified. In 1941 the two draglines that dug in Clearwater County were owned by the J.M.S. Company and the Pilot Dredging Company.[40]

When the federal government outlawed gold mining during World War II, the Pierce area had no dredges in operation; they were already closed down. After the war was over, Clearwater County had almost no dredging activity of any kind. In 1947 Painter and Vincent ran a dragline on Moose and Independence Creeks from June 1 to December 31 and produced 420 fine ounces of gold and 63 ounces of silver. The following year, in a short season, the dredge took out 302 ounces of gold and 43 of silver. From 1950 until 1968, only one other dredge operation was listed in the *Minerals Yearbook*. That occurred in 1961, when the Moose Mountain Mining Company recovered 120 ounces of gold with a dragline in the Moose Creek district.[41]

Notes

1. Hubert Howe Bancroft, "History of Washington, Idaho, and Montana 1845–1889," in *The Works of Hubert Howe Bancroft* (San Francisco: History Company, 1890), 31:427; William S. Greever, *Bonanza West: The Story of the Western Mining Rushes, 1848–1900* (Boise: University of Idaho Press, 1963), 257–64.

2. *Idaho Daily Statesman* (Boise), June 30, 1898.

3. *Salt Lake Mining Review* 5 (May 15, 1903): 16.

4. Idaho State Historical Society, "Early Nez Perce County," Reference Series 334, 1936, 2.

5. George P. Hurst, handwritten notebook, n.d., George Hurst MSS, Box 4, Stanford University Archives, Palo Alto, CA (hereafter George Hurst MSS).

6. *Idaho Daily Statesman* (Boise), November 17, 1903.

7. Ibid., October 19, 1905; *Mineral Resources, 1905*, 237.

8. For these early machines, see Hurst, handwritten notebook, George Hurst MSS, Box 4.

9. *Idaho Daily Statesman* (Boise), January 16, 1906; *Mining Reporter* 52 (July 27, 1905), 95; Albert Nelson Marquis, ed., *The Book of Chicagoans* (Chicago: A. N. Marquis, 1911), 205–6.

10. Layne Gellner Spencer, *And Five Were Hanged and Other Historical Short Stories of Pierce and the Oro Fino Mining District* (Lewiston, ID: Printcraft, 1968), 50; *Mining and Scientific Press* 100 (January 1, 1910): 69.

11. Robert N. Bell, *Ninth Annual Report of the Mineral Industry of Idaho for the Year 1907* (Boise), 140; John Power Hutchins, "Report to the Idaho Company, Limited," copy, November 20, 1908, Charles Janin MSS, Box 31, Huntington Library, San Marino, CA (hereafter Charles Janin MSS).

12. R. H. Bailey to Charles Janin, Spokane, December 18, 1915, Charles Janin MSS, Box 1.

13. *Mining and Scientific Press* 100 (January 1, 1910): 69; *Salt Lake Herald*, February 22, 1909; *Salt Lake Mining Review* 11 (August 15, 1909): 17; Robert N. Bell, *Eleventh Annual Report of the Mining Industry of Idaho for the Year 1909* (Boise), 80.

14. Bell, *Annual Report*; *Idaho Daily Statesman*, August 16, 1909; *Mining and Scientific Press* 100 (January 1, 1910), 69.

15. *Idaho Daily Statesman*, January 18 and July 28, 1911; *Mineral Resources, 1912* (US Geological Survey), 722–23; *Mineral Resources, 1913* (US Geological Survey), 771.

16. Robert N. Bell, *Fourteenth Annual Report of the Mineral Industry of Idaho for the Year 1912* (Boise), 102; Spencer, *And Five Were Hanged*, 51; *Salt Lake Mining Review* 15 (October 30, 1913): 16; *Engineering and Mining Journal* 97 (January 10, 1914): 111.

17. *Mining and Engineering World* 44 (January 1, 1916): 35; *Idaho Daily Statesman* (Boise), January 2, 1916; *Mineral Resources, 1915* (US Geological Survey), 541; *Mineral Resources, 1916* (US Geological Survey), 583; *Mineral Resources, 1917* (US Geological Survey), 476; *Mineral Resources, 1918* (US Geological Survey), 479; *Mineral Resources, 1920* (US Geological Survey), 250–51; *Mineral Resources, 1921* (US Geological Survey), 410; *American Mining Manual, 1919* (Chicago: Mining Manual Company, 1919), 178; *Salt Lake Mining Review* 23 (January 15, 1922): 2.

18. *Bulletin of the American Institute of Mining Engineers* 109 (January 1916): xxx.

19. Spencer, *And Five Were Hanged*, 51.

20. *Engineering and Mining Journal* 98 (November 7, 1914): 853; *Mineral Resources, 1915*, 541; *Mining and Engineering World* 44 (January 1, 1916): 35; *Idaho Daily Statesman* (Boise), January 2, 1916; Sidney Norman, ed., *Northwest Mines Handbook* (Spokane: Northwest Mining Association, 1918) 2:35.

21. *American Mining Manual, 1920* (Chicago: Mining Manual Company, 1920), 182.

22. *Alaska and Northwest Mining Journal* 12 (February 1918): 26.

23. *Mineral Resources, 1920*, 251; *Mineral Resources, 1921*, 410; *Mineral Resources, 1922* (US Geological Survey), 234; Stewart Campbell, *Twenty-Third Annual Report of the Mining Industry of Idaho for the Year 1921* (State Mine Inspector's Report), 49.

24. Spencer, *And Five Were Hanged*, 51; *Engineering and Mining Journal* 114 (August 12, 1922): 298; Stewart Campbell, *Twenty-Fourth Annual Report of the Mining Industry of Idaho for the Year 1922* (State Mine Inspector's Report), 14, 74; *Mineral Resources, 1923* (US Geological Survey), 390.

25. *Mineral Resources, 1926* (US Bureau of Mines), 437; *Mineral Resources, 1929* (US Bureau of Mines), 666.

26. *Spokane Chronicle*, July 5, 1929, clipping, Kimble Northwest History Database, Washington State University Libraries, Digital Collections, http://content.wsulibs.wsu.edu/cdm-clipping (hereafter Kimble Database, WSU).

27. Ibid., July 7, 1932, clipping, Kimble Database, WSU; US Bureau of Mines, *Minerals Yearbook, 1931–32* (Washington, DC: Government Printing Office, 1933), 107.

28. *Spokane Chronicle*, January 4, 1933, clipping, Kimble Database, WSU.

29. *Mining Truth*, April 6, 1933, clipping, Scrapbook, Yuba Manufacturing Company Papers, Meriam Library, Chico State University, Chico, CA; *Spokesman-Review* (Spokane, WA), November 1, 1922, clipping, Kimble Database, WSU.

30. *Spokesman-Review* (Spokane, WA), August 26, 1934, clipping, Kimble Database, WSU.

31. Ibid., April 19, 1935, Kimble Database, WSU; George Edmunds to Charles Janin, Seattle, April 6, 1935, Charles Janin MSS, Box 4, Huntington Library, San Marino, CA.

32. *Spokesman-Review* (Spokane, WA), October 20, 1935, clipping, Kimble Database, WSU.

33. Ibid., June 16, 1935, Kimble Database, WSU.

34. Ibid., August 9, 1937, clipping, Kimble Database, WSU; *Minerals Yearbook, 1936*, 395.

35. *Spokesman-Review* (Spokane, WA), July 26, 1936; January 16, March 20, April 18, May 16, July 25, 1937, all clippings, Kimble Database, WSU; *Spokane Daily Chronicle*, March 19, 1949.

36. *Spokesman-Review* (Spokane, WA), April 18, 1937, clipping, Kimble Database, WSU.

37. Ibid., January 11, 1938; *Minerals Yearbook, 1937* (Washington, DC: Government Printing Office, 1938), 322; *Minerals Yearbook, 1938* (Washington, DC: Government Printing Office, 1938), 351; *Minerals Yearbook, 1939* (Washington, DC: Government Printing Office, 1940), 322.

38. *Spokesman-Review* (Spokane, Wa), July 5, 1936, clipping, Kimble Database, WSU. This article erroneously refers to Rhodes Creek as "Rose Creek."

39. *Spokane Chronicle*, March 5, 1937, clipping, Kimble Database, WSU; William W. Staley, "Mining Activity in the North Fork of the Clearwater River Area," Idaho Bureaus of Mines and Geology, Pamphlet 54 (Moscow: University of Idaho, September 1940), 4; *Minerals Yearbook, 1939*, 326; *Minerals Yearbook Review of 1940* (Washington, DC: Government Printing Office, 1941), 342; *Minerals Yearbook, 1941* (Washington, DC: Government Printing Office, 1942), 348; "Yuba Manufacturing Company List of Mines Using Dredges," typescript copy, 1946, Romanowitz Papers, Rasmuson Library, University of Alaska, Fairbanks, 8.

40. "*R. P. Getty, et al. v. Jett Ross Mines, Inc., et al.*," *Washington Reports*, Dept. 2, 23 (May 22, 1945): 46; *Spokane Chronicle*, April 22 and July 2, 1937, clippings, both in Kimble Database, WSU; *Minerals Yearbook, 1936*, 557; *Minerals Yearbook, 1937*, 322; *Minerals Yearbook, 1938*, 351; *Minerals Yearbook Review of 1940*, 342; *Minerals Yearbook, 1941*, 348; E. D. Gardner and Paul T. Allsman, "Power-Shovel and Dragline Placer Mining," *US Bureau of Mines Information Circular* 7013 (May 1938): 27, 42.

41. *Minerals Yearbook, 1947* (Washington, DC: Government Printing Office, 1948), 1407; *Minerals Yearbook, 1948* (Washington, DC: Government Printing Office, 1949), 1500; *Minerals Yearbook Area Reports, 1961* (Washington, DC: Government Printing Office, 1962), 3:340.

5

Florence

The Florence placers have been described as "the richest short-lived camp in Western America." Located at an altitude of slightly more than 6,000 feet above sea level near the center of a small basin five to eight miles across and surrounded by high snow-capped mountains, the district received annual precipitation of thirty to forty inches, most of it falling as snow and covering the ground for more than six months, often with an accumulation to a depth of ten feet.[1] The site contained a number of small branch creeks and streamlets, on all of which coarse gold was plentiful; on some, the riches were astonishing. In the early days a single yard was said to have yielded $4,500. Peter Bablaine, whose nickname "Baboon" was applied to one of the most productive gulches, left Florence in the spring of 1862 weighted down with seven-five pounds of gold dust. But the boom days were brief; within a few years the bulk of the population had moved on to other diggings, leaving the ground mainly to the industrious Chinese, who washed over the old tailings. No accurate figures of total production are available, only estimates that run from $15 million to $30 million, most of it prior to 1868. By the end of

DOI: 10.5876/9781607324751.c005

the 1880s Florence was practically deserted, but ten years later quartz mines received a new impetus. Many new locations were filed, several mills were erected, and the number of people in the district rose to about 1,000.[2]

A few days before Christmas in 1898 came a brief notice in the Boise *Statesman* that the Florence Gold Mining Dredging Company had been incorporated and had acquired 300 acres on Sand Creek and Meadow Creek about half a mile from the town. The corporate secretary and promoter, R. Gordon Wood of Placerville, announced that a dredge "of the New Zealand type," meaning a bucket-line machine, would be installed the next spring when the roads were open. The company was nominally capitalized at $500,000, divided into 100 shares of $500 each, a radical departure from the standard issue of dollar shares. The company's president was John H. Meyer, also of Placerville, and stockholders were "almost exclusively" from that Boise Basin community and Bucyrus, Ohio—specifically W. H. Pickering and W. S. Blair.[3]

In mid-August 1899 Wood went east to an undisclosed location to buy machinery for "a huge dredge." A few months later Wood and Meyer spent ten days at the placer site, further inspecting the ground and observing the erection of the dredge as it proceeded with a large workforce under the supervision of R. C. Beard. At that time, all parts were at the scene except the massive 100 horsepower boiler, which weighed seven tons, and installation was expected to continue through the winter, with startup in early spring. According to the Lewiston *Tribune,* the plant had the capacity of handling 150,000 cubic yards per month but was expected to work at about half that rate. By that time the company owned about 1,000 acres, 600 of which had been worked in earlier days. Wood made the point that the ground had been recently prospected and the results surpassed earlier estimates of the values. Indeed, he said, with the optimism of a promoter, "It is conservatively estimated that there is sufficient pay ground embraced in the company's holdings to keep 10 dredges steadily running for a period of 20 years."[4]

Transporting the heavy equipment through the mountainous terrain had posed monumental problems. It came from the east by freight train to North Lapwai, the railhead, about a dozen miles from Lewiston. From there it went through Slate Creek country to Florence, 110 miles or so of rugged territory over primitive roads and rough trails, sometimes with creeks and ravines to cross. Experienced packers, freighters, and their teams were employed at exorbitant prices to do the job, which took weeks to complete. The great boiler was hauled by eight teams, at times with men riding long balancing poles on the upper side of the load to prevent overturning. Other parts of the machinery were transported on the backs of pack animals; heavier parts of the equipment were slung from long poles and carried by two horses bearing the front end and two more bringing up the rear.[5]

The exact nature of the dredge is somewhat confusing, but it is clear that it was not a bucket-line dredge, as company officers originally contemplated. In her admirably detailed historical overview of the Florence mining district, Kathryn McKay notes that the Bucyrus Dredge Company installed a $50,000 steam-powered dipper-type dredge in 1899. The shovel was powerful enough to scoop up half a ton of gravel at a time, but in so doing it compacted the gravel and dropped it with such force that it passed through the sluice without disintegrating.[6] Bucyrus manufactured such machines but was not an operating company. It is not certain who built that particular shovel. Bucyrus's records show that it built a number of dipper dredges in 1898 and 1899, but none went to the Florence district.[7] Thus, if it were a Bucyrus product, it would have to have been a used one.

It is also unclear when the dredge was completed and when it started up. Geologist Waldemar Lindgren, who visited the area in 1898 or 1899, reported that in the latter year "a steam dredging enterprise was inaugurated, by means of which it was proposed to wash some areas of low-grade gravels," but he did not indicate a beginning date. Walter Crane, a professor at Columbia School of Mines, wrote in 1908 that in 1899 "a steam gold dredge was in operation on Baboon Creek in the Florence district."[8] Given the company's plans and the ferocity of winters in the area, it seems doubtful, if not impossible, that the new boat commenced digging until the spring of 1900. It is unlikely that there might have been a second dredge that early, although the records are somewhat confusing on this issue. Boise and the Grangeville newspapers mention only the Florence Gold Mining Dredging Company, and both note that its property was on Sand Creek and Meadow Creek.[9] Crane was probably mistaken about the name of the stream on which the new boat operated. A much more recent study indicated that a strip on Sand Creek about 600 feet long and 100 feet wide had been dredged in the early days along the center of the main meadow. Using water backed up in the stream, the company washed some of the low-grade gravels in the deep wet fields and peat deposits.[10]

Once the dredge began digging, there were early favorable reports. Engineer Walter Hovey Hill visited the district on professional business soon after startup and noted that the machine had "been shoveling gravel some three weeks" and that officials on the scene "are very well pleased with the result so far." They were so pleased, Hill said, that "in all probability they will install one or more" additional dredges on their property. In Hill's opinion, the Florence Basin was admirably situated for dredging, and its thousands of acres could provide "plenty of opportunities to work a dozen machines in this field."[11]

But that was not to be; success was short-lived. The Florence Gold Mining Dredging Company soon shut down operations. In the spring of 1900 it was reported that the Baird Brothers' big dredge would employ forty to forty-five men. This was probably the same shovel, although one cannot rule out completely the possibility of a second dredge. Assuming that it was indeed the same machine, in the two partial seasons it operated the dipper produced little—a highly unsatisfactory result for a venture that may have cost a total of $100,000 before it was over. Although an article in the *Idaho County Free Press* in late spring 1902 mentioned talk of starting the machine up again, that never happened.[12]

According to retrospective accounts, the Florence Gold Mining Dredging Company's dredge proved ill-fitted for the task. Old-timers believed it was not a lack of gold that caused failure; it was simply the wrong kind of machinery. In any event, the dredge was abandoned on the banks of the creek. It remained there for years, along with other derelict machinery, including a big dragline. Even though the equipment remained intact on the edge of Sand Creek for years, ownership had passed through many hands, including banks and mortgage companies. Finally, in 1936 various parts were salvaged in an attempt to cobble together a workable dragline machine to recover precious metal.[13] Who was behind this effort and what success it may have had is not clear, but it was probably not very successful. At about the same time, Walter Hovey Hill, a veteran mining engineer who knew the Florence ground well, estimated that dredging had produced less than $50,000 in the district and that at the current value of gold—$35 an ounce—the ground would not return a profit.[14]

How many other attempts were made to dredge in Florence Basin in the interim is uncertain. There is a record that a Mrs. Cullen worked as a cook for the Erbis Mining Company, which ran a dredge of some kind on Sand Creek, but no indication is given as to when or how long it ran. In 1912 the Florence Hydraulic and Dredge Company provided both a bunkhouse and meals for its workers, but information is scanty as to whether the firm lived up to its name or whether its work was limited to hydraulic monitors. A few years later the Wells Company of Denver sent ten men to prospect some low ground, with the idea of applying a dredge if the gravel proved up, which apparently it did not. In the summer of 1923 the Submarine Gold Placer and Quartz Mining Company, a firm originally incorporated to operate an experimental suction dredge on the Salmon River south of Cottonwood, appeared on the Florence scene. With offices in Spokane and Lewiston, it bought placer claims, including the Buttercup and Gold Leaf groups, in Florence meadows for about $30,000, with 40 percent of operating profits to go to the owners until the purchase price had been paid. The company did not issue stock as such; it gave a "Guarantee Certificate" that assured investors a 7 percent return "out of profits." At first, this Seattle firm used a steam shovel, but the machine broke down repeatedly and it dug for only a few days. By 1924, when the company had twenty-eight patented claims, it was using a new steam shovel with caterpillar treads, in conjunction with a dragline, and had a crew housed in an old hotel in New Florence. The shovel was used to strip the ground, which was heavily layered with peat. It must have been a profitable venture because it was operating as late as the 1926 season and worked for some distance—something none of the previous efforts had done.[15]

The gold output of placer mines throughout the West increased during the Great Depression years of the 1930s, and those in Idaho were no exception. One reason was the army of jobless and underemployed who took to the goldfields with pans, rockers, sluice boxes, and other equipment to try their hands at small-scale mining. The historian of this phenomenon called it *The Automobile Gold Rushes*.[16] Often camping in tents with their families, these novice prospectors sought virgin ground or reworked old placers for whatever small gains they might produce. In the summer of 1932 more than 3,000 men were prospecting in the state of Idaho, many of them in Idaho County. But gold mining did not revive in the Florence Basin as much as in the rest of the county. In 1932 and 1933, a total of only $911 was produced in the district.[17]

In the midst of the Depression, a group of mining men in Spokane conducted a one-week basic placer mining school for the unemployed during the spring of 1932. Two thousand people attended and listened to talks by W. W. Staley, professor of mining, and Dean J. W. Finch, both of the University of Idaho. The sponsors thought the school had been "remarkably successful" and followed it with several similar short courses, which attracted thousands more. Staley's primer, *Elementary Methods of Placer Mining*, went through five editions between May 1931 and June 1932. Next, the Northwest Mining Association began to campaign for another idea it believed might get large numbers of miners back to work. The federal government should pursue dredging and hydraulicking on a large scale on properties where there was known to be considerable gold but not enough to attract private capital. Vast areas of placer ground containing values of five to seven cents per yard had been examined and rejected by a number of companies, which would be willing to share the test results with the government, it was argued. The federal government would pay the cost of the mining but in so doing would create new jobs and add to the world's supply of gold. A proposal to this effect was submitted to the Northwest Mining Association, whose president, Robert T. Banks, promptly appointed a three-man committee to draft resolutions to be sent to Washington, DC.[18] No action was taken by the federal government.

A bright spot appeared at the end of January 1934, when President Franklin D. Roosevelt devalued the dollar and set the price of gold at $35 an ounce, up from $20.67—an obvious boon to the gold mining industry. At about the same time, new technology—the dragline and floating washing plant—was lowering costs and adding flexibility and was quickly adopted in the Florence area. The Jones family was one example. The family owned a two-thirds interest in the Homestake group of six claims extending up Sand Creek from near the mouth of White Sand Creek. In 1935 J. F. (Frank) Jones and Leo Sitar organized the Homestake Placers Company to work both the Homestake group and the Buttercup group nearby. The firm's machinery included the old Submarine Company shovel, retrofitted with a log boom and dragline, with a home-built floating washing plant and a small Cat for clearing the ground. With that equipment, in the summers of 1935 and 1936 Jones and Sitar mined a strip at the lower end of the Homestake ground. Reports differ as to the size of the strip and the value of the gold produced. H. S. Lorain and O. S. Metzger say that the already dredged area was about 1,000 feet long and 30 feet wide and that results were positive, with gravel running more than 40 cents per yard. Geologist John C. Reed maintains that the area dug was 210 feet by 40 feet, and the total value ran about $1,200, for an average yield of 16 cents a yard on ground that required peat removal and 28.5 cents on ground that did not. In any event, Jones and his partner closed down their operation by September and moved their machinery.[19]

In the early 1930s Otto Egloff had some success with a small rocker on Healy Creek on worked-over ground covered with tailings. In 1936 a Boise outfit, the Granite Creek Dredging Company, headed by W. L. Smith, transferred dragline and a floating washing plant from the Boise Basin and began digging on property belonging to Egloff, eventually expanding to ground owned by Jack Hardin. With Smith as manager and a crew of sixteen men in two shifts, the company worked through five feet of overburden to reach three feet of paydirt on bedrock. Again, accounts differ on the degree of success. One report insists that the rig's initial season was a winner; another argues that it failed and that the property was in litigation until 1938. An Idaho newspaper item in July 1937 indicates that all was going well on the Egloff property: the company had full crews and "a complete camp and reports successful operations." A shortage of water prompted a closedown early that year. Another mining note from a Spokane paper ten months later heralds the opening of the 1938 season for the Granite Creek Dredging Company.[20]

During 1937 and 1938, Walter Jones used a small gasoline shovel, a bulldozer, and a movable washing plant to handle about 80,000 cubic feet on branches of Sand Creek and on Meadow Creek. It was no doubt the Jones operation that enabled the *Minerals Yearbook* for 1938 to note that almost all of the output of the Florence district was placer gold and that the largest producer was a dragline rig on Meadow Creek. Subsequently, a year later a dragline and floating washer also dug with some success on the Sterling ground on Sand Creek.[21]

While the early 1940s were banner years for gold placer mining in much of Idaho County, for the Florence Basin they were lean times, with gold yields so small and sporadic as to be unnoticed in the annual *Minerals Yearbooks*. No dredges were working. Then came World War II, and what had been record seasons of production in the state ceased abruptly.

In October 1942 the federal government deemed gold mining unessential and closed down operations during World War II. Thus, production for 1943 was nearly 68 percent less than that of the previous year. No Idaho dredges operated that year; the total recovery recorded in the Elk City district was ten ounces. The output of placer gold in the state for 1944 was the smallest since 1862. Some gold was still produced, but it was largely a by-product of mining other metals. When the War Production Board finally rescinded its order banning gold mining on July 1, 1945, only one bucket dredge, that of the Idaho-Canadian Dredging Company at Idaho City, was able to operate during the final three-and-a-half months of the year. In Idaho County, total recovery was fifty-four ounces, split among seven districts.

Gold production elsewhere in central Idaho gradually increased over the next five years, but not in the Florence district. Not until 1958 was there any indication of a serious renewal of dredge activity in that area. This came with the incorporation of Idaho Mining and Milling, Inc., headed by Philip Jungert of Lewiston, with his brother, Marion, as secretary-treasurer. The company's nominal capital was listed as 45 million shares at 2 cents each.[22] The firm had plans to develop a substantial tract of ground and ultimately owned the rights to more than forty miles along Sand and Meadow Creeks and their tributaries—property largely untouched by heavy equipment. Testing in 1957 was encouraging and showed some pockets with gold values running as high as $3.00 a yard, but the ground on which operations would begin averaged just $1.37 per yard—still a handsome figure, according to information the Jungerts sent to "Dear Co-Owner" investors. In the interim, the company had built a sawmill, a general headquarters, a machine storage building, and an airport—the latter deemed vital in the isolated district. "Stripped ground ahead of the dredge can be leveled for a runway," the brothers explained. "As the dredge eats its way through the airstrip the dredge tailings can be leveled for a permanent air field."[23]

In the spring of 1960 they announced that a dredge had been acquired in Warren and that Leo Sutton and Sons had agreed to dismantle the machine, haul it to the Florence site, and work with the firm's crew in assembling and erecting it again.[24] At about the same time, Idaho Mining reported to the US Bureau of Mines that it was readying a dredge, but it did not become operational until the following year. The machine was a diesel-electric bucket-line dredge, one of three Yuba boats the company is said to have owned by 1962. Starting at the mouth of Sand Creek, its dredge worked up Meadow Creek, including its west and main forks.[25] On its maiden run in

1961, the dredge had recovered 580 ounces of gold by mid-October. During a second short season, it produced 122 ounces of gold and 48 ounces of silver before it was forced to close down by siltation the machine's operation caused in the streams.[26]

This was the most serious of the company's problems, but there were plenty of lesser ones. Water was in short supply, harsh winter weather made for a short working season, and layered peat on some of the ground made progress difficult. There was difficulty saving the gold because of the quicksilver left in some of the ground by earlier miners using more primitive methods. Where the company had hoped to also reclaim important heavy minerals and rare earth from the black sands remaining from previous placer operations, the district's inaccessibility made it difficult to reach distant concentrating plants. In any event, it is not surprising that in 1963 Idaho Mining and Milling, Inc., the largest gold producer in the county at the time, announced plans to end its operations.[27] Later, a photograph of the firm's boat billed it as the last bucket dredge to operate in Idaho. In another district, Clair Johnson's dragline and floating washing plant worked along the west side of the American River northeast of Elk City as late as July 6, 1981.[28]

Notes

1. Muriel Sibell Wolle, *The Bonanza Trail* (Bloomington: Indiana University Press, 1953), 222; John C. Reed, "Geology and Ore Deposits of the Florence Mining District, Idaho County, Idaho," Idaho Bureau of Mines and Geology Pamphlet 46, Moscow, ID, 2–3, 5.

2. Waldemar Lindgren, "The Gold and Silver Veins of Silver City, De Lamar, and Other Mining Districts in Idaho," in *Twentieth Annual Report of the United States Geological Survey to the Secretary of the Interior 1898–1899*, edited by Waldemar Lindgren, part 3 (Washington, DC: Government Printing Office, 1900), 233–35; Cornelius J. Brosnan, *History of the State of Idaho* (New York: Charles Scribner's Sons, 1918), 91–92; *Irrigation Age* 12 (March 1898): 172.

3. *Idaho Daily Statesman* (Boise), December 20, 1898.

4. Ibid., August 14, October 17, November 19, 1899; *Lewiston (Idaho) Morning Tribune*, August 30, 1936, clipping, Wallis and Marilyn Kimble Northwest History Database, Washington State University, Pullman (hereafter Kimble Database, WSU).

5. *Lewiston (Idaho) Morning Tribune*, August 30, 1936, clipping, Kimble Database, WSU.

6. Kathryn L. McKay, "Gold for the Taking: Historical Overview of the Florence Mining District, Idaho County, Idaho" (Grangeville, ID: Nez Perce National Forest, 1998), 141.

7. Bucyrus Company, "List of Dredges, 199975 A & B," possession of the author.

8. Lindgren, "Gold and Silver Veins of Silver City," 235; Walter R. Crane, *Gold and Silver* (New York: John Wiley and Sons, 1908), 74.

9. *Idaho Daily Statesman* (Boise), December 20, 1898; *Idaho County Free Press* (Grangeville), December 30, 1898, quoted in Mary Alfreda Elsensohn, *Pioneer Days in Idaho County* (Caldwell, ID: Caxton, 1957), 1:96–97.

10. H. S. Lorain and O. H. Metzger, "Reconnaissance of Placer Mining Districts in Idaho County, Idaho," *US Bureau of Mines Information Circular* 7023 (June 1938): 48; G. J. Orris and J. D. Bliss, "Geologic and Grade-Volume Data on 330 Gold Placer Reports," US Geological Survey Open-File Report, 85–213, 87; McKay, "Gold for the Taking," 141.

11. *Idaho Daily Statesman* (Boise), April 26, 1900.

12. McKay, "Gold for the Taking," 141; *Idaho County Free Press* (Grangeville), spring 1902.

13. Robert C. Bailey, *River of No Return* (Lewiston, ID: Bailey-Blake, 1935), 99; *Lewiston (Idaho) Morning Tribune*, August 30, 1936, clipping, Kimble Database, WSU; US Forest Service, Nez Perce National Forest, "Florence Tells Her Secrets," http://www.fs.usda.gov/Internet/FSE_DOCUMENTS/fsm91_055709.pdf.

14. Elsensohn, *Pioneer Days in Idaho County* 1:73, 75.

15. McKay, "Gold for the Taking," 141, 145, 147, 177; Submarine Gold Placer and Quartz Mining Company Guarantee Certificate, issued to Ida and R. G. Metcalf, December 9, 1925, signed by A.B.C. Davis, secretary, and Lewis Williams, president.

16. For an excellent detailed treatment of this movement, see Charles Wallace Miller Jr., *The Automobile Gold Rushes and Depression Era Mining* (Moscow: University of Idaho Press, 1998).

17. McKay, "Gold for the Taking," 147, 150.

18. William Wesley Staley, *Elementary Methods of Placer Mining* (Moscow: University of Idaho, 1931); *Spokane Chronicle*, June 14, 1932, clipping, Kimble Database, WSU; Richard C. Waldbauer, *Grubstaking the Palouse: Gold Mining in the Hoodoo Mountains of North Idaho, 1860–1950* (Pullman: Washington State University Press, 1986), 33–34.

19. Lorain and Metzger, "Reconnaissance," 47–48; Reed, "Geology and Ore Deposits of the Florence Mining District," 35; *Lewiston (Idaho) Morning Tribune*, September 28, 1936, clipping, Kimble Database, WSU.

20. McKay, "Gold for the Taking," 177–78; *Lewiston (Idaho) Morning Tribune*, September 28, 1936; *Kellogg (ID) Evening News*, July 16, 1937; *Spokane Chronicle*, May 6, 1938, all clippings, Kimble Database, WSU.

21. McKay, "Gold for the Taking," 151, 170, 178; *Minerals Yearbook, 1938*, 352; *Minerals Yearbook, 1939*, 328.

22. See "Idaho Mining and Milling, Inc.," stock certificate, issued to J. A. Hogle and Co., August 28, 1962.

23. Idaho Mining and Milling, Inc., general letter to "Co-Owner" (no name, March 1960), copy, in Supervisor's Office, Headquarters, Nez Perce National Forest, Grangeville, ID; McKay, "Gold for the Taking," 180.

24. Idaho Mining and Milling, Inc., general letter to "Co-Owner" (no name, March 1960), copy, in Supervisor's Office, Headquarters, Nez Perce National Forest.

25. *Minerals Yearbook, 1960* (US Geological Survey), 3:339; McKay, "Gold for the Taking," 181.

26. McKay, "Gold for the Taking," 180–81; *Minerals Yearbook Area Reports, 1962* (US Geological Survey), 3:357.

27. *Minerals Yearbook Area Reports: Domestic 1963* (US Geological Survey), 3:352; McKay, "Gold for the Taking," 180–81.

28. Sarah Elizabeth Purdy, "Analysis of Dredge Tailing Pile Patterns: Applications for Historical Archaeological Research," MA thesis, Oregon State University, Corvallis, June 7, 2007, 54, 56.

6

Elk City and Newsome

Located in the center of Idaho County, fifty mile east of the present city of Grangeville in the southern foothills of the Bitterroot Mountains, the Elk City mining district developed in a manner similar to the state's other high placer areas. The main town and center for mining in the district, Elk City, was set in a basin amid moderate hills at an elevation of 4,200 feet above sea level. Elk Creek, the American River, and the Red River flowed through that basin and joined west of town to form the South Fork of the Clearwater. From the south, the Crooked River ran northward and fed into the South Fork still farther to the west. Lesser tributaries of all these streams provided ample placer ground to stimulate the imagination of many gold seekers. The camp itself went back to 1861 and was a lively and productive place for a few years, during which several million dollars worth of gold was taken out.

The Elk Creek basin was in the form of oval depressions about eight miles long and three miles wide, filled with detritus. In addition, in numerous parts of the district there were placers of what geologists called the "re-concentrated type," where erosion had removed much of the high-level

DOI: 10.5876/9781607324751.c006

gravel, leaving behind gold in the streambeds or in the underlying bedrock. While bedrock at the bottom of the streams was usually too deep to be reached readily, most early mining was limited to the high bars and gravel terraces where the paydirt was within a few feet of the surface. Such "skim" diggings produced heavy returns, but after a few good seasons they began to play out, and the Florence excitement and new strikes in Montana attracted miners to more profitable fields. As was often the case elsewhere, for about a dozen years after 1872 the Chinese dominated mining in the district, with renewed activity by whites increasing gradually in the mid-1880s and picking up momentum by the turn of the century. Hydraulic mining was already established. Thoughts began to turn to new technology.[1]

Evidence indicates that the first dredge came to the Elk City area in 1891, when gold dredges were still in the unproven nascent stage. One C. D. Gavins of Chicago was responsible. What kind of rig it was and how long it worked is uncertain. The machinery was packed in on the backs of mules, and, once assembled, it dug on the banks and bed of the Red River for some time. In 1892 Gavins bought the Reynolds and Mooney ranches on Red River Meadow and could thus follow the pay streak up to the flat. It was reported that Gavins named the small settlement of Raymond, near where the dredge worked, for the son of a financial backer in Illinois. Ultimately, the machine proved too light and had to be abandoned. After a few years, Gavins turned up in the Boise area, where he was active as a mine promoter in the mid- and late 1890s.[2]

A newspaper story in early 1899 indicated that the Red River Mining Company had worked its property 9 miles from Elk City "five or six years ago with the old-fashioned bucket dredge, but the appliances did not work to any great success." The account went on to say that the Butters brothers had bought 1,280 acres of ground on the river meadows from the company the previous year, had thoroughly prospected it, and were so satisfied that they had ordered a Bucyrus bucket dredge, expected to cost $35,000. It was anticipated that the new machine would be operational by April 1, and an electric light plant had been installed to enable it to run day and night. W. H. Butters was there from Spokane to supervise the transportation of more than 80,000 pounds of freight—equipment and supplies that had accumulated at the railhead in Lewiston, 140 miles away.[3] A good deal of the route was over rugged mountainous terrain, and heavy machinery required big wagons and four- or six-horse teams. Lewiston was 756 feet above sea level; the road climbed to an elevation of about 6,500 feet near the summit before eventually dropping several thousand feet to the Elk Creek mines.[4] In any event, after four days on the road, thirteen teams arrived in late February hauling the Red River dredge.[5]

When the dredge arrived and what kind it was were not reported. If it was a Bucyrus, it was not a new one. The Bucyrus Company's comprehensive listing does not show any machine constructed for the Elk City area in the late 1890s or early 1900s. A detailed chronicle on the history of Idaho County published much later in the century mentions that the Butters brothers rebuilt the old dredge but that it, too, was abandoned after a short time.[6] In late 1900, when the Central Idaho Gold Dredging Company was organized with nominal capital of $50,000, C. E. Butters was listed as the Elk City agent,[7] although no further mention is made of a firm with that name. In early 1901 the Boise *Statesman* carried a personal item to the effect that Gardner I. Porter, former superintendent of the American Hill placer at Elk City, would take charge of the Butler brothers' dredging operation as soon as repairs on the boat were completed and the spring thaw was assured—probably in about two weeks.[8] This leaves open the possibility of an editorial error in names. Could "Butler brothers" perchance have been intended to mean "Butters brothers"?

In 1904 the Idaho Mining Company had a dredge working the old tailings and the gravel underneath near the little town of Raymond on French Creek, a feeder of the Red River. The same outfit was also running a large hydraulic operation at American Hill, about a mile southeast of Elk City.[9] How long this boat may have been digging is not clear. There was never any mention of it in the federal government's annual *Mineral Resources* series. According to Mary Elsensohn, interestingly, in 1941 the brush-covered remains of an old dredge could still be seen in French Gulch off the Red River.[10]

Not until 1909 did another gold dredge catch the eye of local residents. A key player in its appearance was Patrick J. Jennings of Portland, who at different times acted as middleman, promoter, and manager. Jennings had mining investments in Lane and Douglas Counties, Oregon, and the Coeur d'Alenes and was a vice president of the Oregon Mining Stock Exchange, as well as one of the founders of the Elk City Bank.[11] In late 1908 he had returned from a Chicago business trip to Elk City, where he was said to have an interest in dredging property.[12] Next came word that Jennings was operating the Gold

Seal Placer near Elk City and that he would install a large dredge during the coming season; indeed, according to the press, one had been ordered from Milwaukee. Subsequently, it was announced that Jennings had sold the property to the Chicago Dredging Company and bought 180 acres of the Golden Scale and Baboon Placers, located on the American River two miles from Elk City. Sometime in mid-1909 he organized the Elk City Dredging Company, with himself as president and capital stock of $600,000, in shares of $100 each. According to the Idaho inspector of mines, the company was actually controlled by Cleveland investors.

Next, Jennings or his company purchased the Gold Seal Extension from J. B. Henniger and John Massam, with final cash payment in late May. Dr. James P. Kimball, a former director of the US Mint and a distinguished mining engineer in his own right, was brought in to examine the property, although he seems to have had some private financial interest involved. In testing the new ground, Kimball "proved all that the early miners claimed for it"; indeed, the good doctor found prospecting values of more than 70 cents a yard. It is therefore no surprise that agents of well-known engineer and mine owner Fred W. Bradley tried to acquire it but found it tied by the Jennings group, who were convinced that dredging was the only way to work it. In the meantime, Jennings had not ordered a dredge from south Milwaukee, as rumored; instead, he had contracted with Isaac B. Hammond of the Hammond Manufacturing Company in Portland for construction of two bucket-line machines, When Hammond, who was finishing up a dredge at Gold Center in Grant County, Oregon, arrived on the scene, he ordered his shipbuilders to bring along a sawmill and a planer to be installed on upper Elk Creek, close to the yellow pine belt.[13]

At this time, all supplies and equipment had to be hauled by wagon from the railhead, which was no longer Lewiston but States, fifty-eight miles from Elk City. Transporting freight was still both arduous and expensive; as late as 1914 it cost two cents a pound.[14] What a Hammond dredge or two might weigh is uncertain, but the Hammond sawmill weighed 47,000 pounds when it arrived by rail at States and was ready to be loaded onto wagons for the rest of its journey.[15]

The two boats were completed by mid-November 1909, both under the supervision of Patrick J. Jennings. One, digging under the flag of the Elk City Dredging Company, was to operate at the confluence of Elk Creek and the

American River. The other was to work the placers on Little Elk Creek, three miles north of town, under the name of the Jennings Dredging Company.[16] Jennings was manager of the Chicago Dredging Company in early 1909 in the same area before the dredges were built.[17] Likely, the Jennings Dredging Company was a restructuring of the older firm.

Although Jennings had hoped to get a month or so of work in before winter struck, that seemed to have proved impossible. At the end of November the Little Elk Creek boat was digging day and night, thanks to electric lights, and had discovered a false bedrock of blue clay on its ground, beyond which the early miners had not penetrated. When reached by the Jennings Dredging Company machine, the true bedrock proved much richer than previous finds. The local *Mining Review* gives the impression that the second Hammond dredge was operative before the end of the 1909 season. If so, it may have been simply a brief shakedown cruise to test and find the bugs in the machinery. Other sources suggest that only one boat was active, and only for a short time, before cold weather shut it down.[18]

By early 1910 the mining press was referring to the Elk City Dredging Company as the Idaho Dredging Company. At the same time, the former firm was hawking its stock and its virtues in the advertising section of *Sunset,* the publication of the Southern Pacific Passenger Department.[19] Whether the corporate name actually changed is a question that remains unanswered.

According to the US Bureau of Mines, in 1910 the two boats "were operated but little," and for the next four years they did not run at all. By 1916 the Jennings Dredging Company was included in the eight companies in the state listed as recently having closed down their dredges.[20]

Whether the difficulties that prevented the two outfits from working were financial, mechanical, or otherwise is not clear. Perhaps there was a scare in 1910 for all owners of placer property in the district when the Home Placer, owned by Mrs. J. A. Parr of Elk City, was reclassified as farmland by the federal land office in Lewiston. The higher section of Parr's ground had been thoroughly worked over earlier, but the lower portion had been held in reserve for years in the belief that it was valuable dredge property—an opinion buttressed by the fact that one of the two new companies nearby had taken out values of more $1,000 on its first cleanup. The idea that it should now be handed over to a "claim jumper" who filed on it for a homestead was anathema to the little mining community, to say the least. Parr may have won her appeal to Washington; she sold off eight of her placer claims in the same area about two years later.[21] But for whatever reason, the Elk City Hammond dredges worked only briefly in their lifespan and must have been an abject disappointment to their proprietors.

In the following years, there were several times when it seemed possible that new dredges might be introduced. In the spring of 1914, for example, articles of incorporation were drawn up and registered for the Elk City Gold Dredging Company, which was to conduct a general mining and milling business, with Elk City as its headquarters. With nominal capital of $150,000 in $1 shares, its trustees were a group of Washington state investors—W. A. Luther and M. A. Lungo, both of Spokane, H. A. Renand of Sprague, and A. C. Harmon of Revier.[22] Probably running a well-used machine, the company dug on the American River for about ninety days in 1915, when most of the county's meager output of $7,084 was attributed to it. The following spring it announced plans to install a new dredge "and other placer machinery" during the summer, with C. D. Bussong in charge. The only further mention of this boat was on New Year's Day in 1916, when it, too, was listed as "closed down."[23]

In 1919 rumors were circulating that thousands of acres were being taken under option by the Guggenheim's Yukon Gold Company, operating along Prichard Creek in the Coeur d'Alenes, as well as in California, Alaska, and the Klondike. The ground in question was along the American River and Elk Creek and was said to include a strip two miles wide and five miles long. Drills had been ordered for testing.[24] Perhaps property was examined, but the Guggenheims did not come to Elk Creek.

Prior to World War I, Idaho's annual production of gold averaged about $1.2 million. During the war demand increased for lead, silver, zinc, and copper, with prices of these metals rising sharply while output correspondingly rose by more than 100 percent in 1917. At the same time, the production of gold fell to $804,809, a drop of over 60 percent. When peace came, so did hard times for western mining. Wages, freight rates, and the price of commodities and machinery all went up, and an excess profits tax cut into corporate profits and had a demoralizing effect. By 1921 some relief was on its way. The excess profits tax was repealed, substantial reductions were made in rail freight rates, and more normal conditions prevailed, but these were not boom years for the mineral industry in general.[25]

Another problem Idaho inspector of mines Stewart Campbell pointed out was that mining in the state was being hurt by illegitimate promotions and mismanagement. Of the two, illicit promotions, with their fraudulent literature, were doing the greater harm. Although usually short-lived, such dealings undermined public confidence, especially when the investor did not condemn the promoter but instead went after the state for making the dealings possible.[26] By nature of its ability to test and make reasonable predictions, gold dredging, after its early days, experienced far less of this kind of chicanery than did hard-rock mining.

The reality that mining in Idaho and in the West in general was in the doldrums in this period was well-known. S. F. Hunt—prospector, geologist, philosopher, and the man who discovered Nevada's rich Rio Tinto copper deposit in 1919—recorded his remarks on the state of the metal mining industry half a dozen years later. "The old processes of mining and dealing in mines gathered their obsolete trappings about them in 1910 and silently stole over the Great Divide. Nobody seems to know who killed Cock Robin," he said. "By the great majority who have wondered and worried over the demise, the world war and fear of others to come, revolutions and internecine strife in Africa, Mexico and China, bolshevism, political and social unrest, income taxes and tax-exempt securities, 'blue sky' laws, steam-shovels, and even marriage, the giving in marriage and Reno, are singly or collectively dragged in as the accused assassins."[27]

Not so, Hunt continued. "Evolution killed Cock Robin." Mining in the nineteenth century "was characterized by booms, stampedes, gun play, and active touting and operating of numerous small companies on mostly hot air or high-grade near-surface ore deposits." The transition to twentieth-century mining practice exhibited hesitancy—"cold feet" and a "rapid decrease in the number of operating companies on lower grade, larger and deeper ore deposits." Twentieth-century companies came with new methods, much larger holdings, and greater financial resources. Emphasis was now on efficient mass production, with lower speculative risks, losses, and operating costs. While deploring the lack of exact data on the subject, Hunt pointed out that in 1890 there had been 6,000 precious metal mines in the country, while in 1925 there were no more than 1,250, illustrating the new trend. Of course, there was the question of how many of these companies were actually operational. Back in 1912, the Idaho secretary of state had made a discovery that surprised him: of 1,500 mining companies of all kinds operating under the authority of his office, only 21 were listed as active and had paid corporation taxes for 1913. Nonproductive firms merely filed an annual statement.[28]

In the fall of 1929 a comprehensive exploration of ground near Elk City was completed by the 96 Mining Company. Two drills had been used on Elk Creek and for seven miles on the American River, said a spokesman for the firm. Some of the property had been mined, but none had previously been systematically prospected. While not wide, the pay streak averaged thirty-five cents to forty cents per yard, which would be profitable should a dredge be brought in. The Hinton brothers of Tacoma and their cohorts were said to be the moving spirits behind the move.[29] Apparently, the time was not yet ripe, and no action was taken.

Subsequently, in the mid-1930s, after the price of gold had been raised to thirty-five dollars an ounce, more attention was paid to mining the precious metals and the application of advanced technology to the well-worked placer fields along the South Fork of the Clearwater and its branches. Thus, it might be said that the region's most important dredging era started after 1935 and extended into the 1940s and 1950s.[30]

"Dredges Invade Elk City Region" was a flashy headline in the *Spokane Chronicle* in the fall of 1937. Actually, those particular words were not references to the old-style bucket dredges but to S. D. Alexander and associates from Tacoma and San Francisco, who brought in a diesel-powered dragline plant with a two-yard bucket and washing equipment mounted on a barge.[31] When installed on the South Fork between the mouths of Santiam and Rabbit Creeks, it represented an investment of about $75,000 and worked for forty-five days, "certainly tearing up that river bed," as one observer put it, before winter struck.[32] The next spring the Alexander group was bought out by the Western Gold Corporation of Boise, a firm recently organized to take over several important enterprises, including the Idaho Gold Dredging Company.[33]

Prior to 1938, little dredging had been done on the American River, apart from the ninety-day run of the Elk City Gold Dredging Company in 1915. Its lower half mile was a steep-walled constricted canyon, and even where the valley opened up, there were too many large boulders and the stream was too narrow for the typical bucket-line dredge to operate profitably. Ultimately, much of the American River would be dredged, but that would be accomplished by draglines using floating washing plants. One was owned by the American River Mining Company, which worked from 1939 to 1942 and which handled 246,500 cubic yards of gravel in 1939 and was credited with producing much of the area's gain that season. The firm was owned by president-manager G. B. Peterson, and its vice president–agent was Walter B. Peterson. Its 1½-yard dragline employed twelve men regularly, with additional hands part of the year for prospecting and testing. Replacement cost of its Bodison washing plant and diesel-powered dragline was listed at $60,000. In 1939 the washing plant was destroyed by fire, after which the firm moved its dragline and installed new washing equipment on the upper South Fork of the Clearwater River. The machine operated there until it was closed down by the government ban in 1942, but it never appeared after the war. An outfit called the Cal-Idaho Mining Company was also active briefly in the Elk City district in 1939.[34]

Another dragline working on the American River beginning at about the same time was owned by a Seattle firm, the Tyee Mining Company. Its machine dug for two years before the ban, returned for a five-month season in 1946, then continued until November 1950, when its spokesman announced that it was shutting down operations after nearly seven years of digging. But the company hadn't given up; it had found new ground on the Red River, a few miles south of Elk City, and was busy again in 1951. Tyee worked two more years before calling it quits for good. Before it was finished, manager Kenneth Coates had become curious about the coarse, heavy black sands discarded each time the jigs were cleaned of gold. So he sent samples to the US Bureau of Mines and the US Atomic Energy Commission. The material turned out to be uranium-bearing, not profitable to recover under current economic conditions.[35] Still another dragline operation on the same river was that of Clair Johnson, which worked the west side of the stream northeast of Elk City. Using what has been called "a dragline-dredgeboat combination," Johnson was digging sporadically between 1961 and July 1981.[36]

A tributary of the South Fork of the Clearwater located ten miles south of Elk City, Newsome Creek, was heavily dredged from the mouth of Nugget Creek almost to the mouth of Radcliff Creek in the late 1930s. The Newsome mining district included the upper portion of that creek and its feeders north of the mouth of Bear Creek. The district abutted and shared part of the Ten Mile mining district in nearby Elk City. As in most early booms, the Chinese had replaced the whites by 1880, but white interest was renewed a decade or so later and continued into the twentieth century, with a number of large placer operations that had waned by 1920. No dredging had been done on Newsome Creek prior to the mid-1930s.[37]

Changes began in 1936, when veteran dredge operators in the Warren area, Fred Baumhoff and Al Fisher of the Idaho Gold Dredging Company, bought out a competitor, the Warren Creek Company. The two had been acquiring placer ground in the Elk City area for some time. When they took over the assets of the Warren Creek Company, those assets also included placer ground on Newsome Creek, in the mining district adjacent to the Elk City district. Baumhoff and Fisher immediately announced a challenging plan for mining the property. Their large dredge at Warren would be dismantled at once and be put into operation on Newsome Creek within sixty days. In the meantime, with Cat and bulldozer, Fisher would reconstruct about eleven miles of the old road from the Elk City highway to the old Newsome house.[38]

For some reason, this never happened. Instead, in September 1937 the Newsome Creek Mining Company of California put a crew of twenty men to work under engineer L. N. Ellis building a three-yard, two–diesel engine dragline and a Bodinson washing plant on the site. The outfit was owned by Joe Ferris and Johnny Marchbank, former operators of the Warren Creek Dredging Company. Ferris was president of the firm and, along with Marchbank, owned a substantial share. The two men also owned dragline dredges in California and another in the Canyon City district of Oregon, on the John Day River, near the town of the same name. Marchbank was the former proprietor of a hurdy-gurdy house in Lewiston, as well as the owner of a large racetrack close to San Francisco, and he was regarded as a "big spender" on the Pacific Coast.[39]

At the end of two very successful years, the property of the Newsome Creek Mining Company was taken over by the Ferris Mining Company, which after a partial season ran out of profitable ground. Ferris and Marchbank moved its plant to Unity, Oregon.[40]

Several bucket dredges were active in the Elk City–Orogrande region during the boom that transpired after the price of gold was officially raised in 1934. The Mount Vernon Gold Mining Company dredge, another machine that belonged to H & H Mines, Inc., and was owned by the Warren Dredging Company, also played a role in high-level placer mining in Idaho County in the decade and a half that followed.

From 1935 to 1937, David E. Henderson of Mount Vernon, Washington, operated a small bucket-line dredge six miles south of Elk City on the upper end of Deadwood Creek, a tributary of the Crooked River. The machine, which he named after his hometown, had capacity of two cubic feet, was of standard design, and was able to dig twenty feet below the water and ten feet above. The all-steel boat was powered by two Caterpillar diesel electric generators and had been built in 1935 at a cost of $37,350 by Washington Iron Works in Seattle, a firm that had revolutionized lumbering with its steam skidder but was a relative newcomer to the dredge field.[41] One of its two generators provided seventy-five horsepower for the ladder and screen, the other fifty horsepower for pumps and other equipment, including lights fore and aft for night work. Fuel cost eleven cents a gallon or ten dollars a day for the dredge, plus an additional one dollar a day for the Cat used to clear the ground. For round-the-clock operation, a crew of a dozen was required: three winchmen, three engineers, three deck hands, and from one to three shoremen. Deadwood Creek produced coarse gold that was easily recovered. The working season usually ran from May through November, although it was possible to begin earlier and dig longer but at a much less efficient level. One drawback was that in extremely low water it was necessary to bring additional water from Relief Creek as a supplement. In 1935 and 1936, the Mount Vernon dredge "was by far the largest producer of gold" in the Elk City district.[42]

Late in the summer of 1937, Dave Henderson began testing on Crooked River proper near the Orogrande lode mine in the Orogrande district, a few miles southwest of Elk City township, in hopes of finding new dredge-able ground. Results were encouraging, and the boat was moved from Deadwood Creek and put to work in 1938 on property leased from Orogrande-Frisco Gold Mines, Inc., on the Crooked River bars above Relief Creek. By contract, Henderson was to pay Orogrande-Frisco a 10 percent royalty on gross profits from processed gravel that averaged between twenty cents and forty cents per yard. By 1939 the Mount Vernon dredge was a stalwart producer in the district, and its output continued to be high. In 1941, in a nine-month season when it recovered 2,000 ounces of gold and 430 ounces of silver, it led Idaho County, but the following year it produced 955 ounces of gold and 204 ounces of silver, as it dug only from March 15 to August 8, 1942.[43] Also on August 8 the War Production Board issued Order L-208, deeming gold mining nonessential to the war effort and ordering all such operations, both placer and lode, halted for the duration.[44]

The ban on gold mining during World War II was obviously a hardship for the dredging industry. Only a few boats could continue to operate to produce minerals deemed critical to the war effort, and none of them were in Idaho. In an era of labor shortage, dredge personnel might be drawn into one of the services or readily find jobs in one of the many wartime manufacturing plants, but recovery would be slow after the conflict ended. Gold mining communities like Warren felt the economic pinch particularly keenly. One mining editor reported in early 1943 that at least four flourishing gold mining camps had been reduced to "ghost towns" in Idaho County because of the government ban and had failed to hold elections because nobody was there to vote. In 1938, when E. T. Fisher stepped in by creating a fund to help keep the Elk City school afloat, the town had one store and a population of about 160, with 28 schoolchildren. With the closing down of the mines and dredges, the number of people dwindled to a handful; Sister Mary Alfreda Elsensohn said the population was down to three old prospectors.[45]

Companies and western states damned the closings from time to time, and legislatures passed annulment resolutions. In the spring of 1943 thirty senators from the West, including those from Idaho, petitioned the War Production Board for "immediate recession or modification of the order"—all to no avail.[46]

Not only were dredge personnel affected, but so were the plants that made the equipment. Factories quickly tailored their products to fit the war effort. Yuba Manufacturing Company built 155-millimeter howitzer field guns. Bucyrus-Erie turned out "heavy transit wagons"—mounts for the same caliber howitzer and eight-inch guns—plus general purpose excavators for the military. Bucyrus also constructed the Dozershovel, which became standard among island-hopping US Marines in the Pacific campaigns. This was a machine with caterpillar tread and interchangeable blades and buckets, which enabled it to be used as either a bulldozer or a tractor shovel.[47]

All of the mining companies had lost a good deal of production during the closure, and it was inevitable that they would attempt to recoup their losses by filing lawsuits against the government. Almost as soon as Order L-208 had been issued, western members of Congress talked about federal funding to help the owners, but no legislation was forthcoming. Representative Clair

Engles of California began a campaign in early 1945 to compensate owners but got nowhere. Once the war was over, Engles in the House of Representatives and western senators jointly introduced bills to provide relief—bills President Harry Truman soundly condemned and the critical *New York Herald Tribune* called "legislative gold digging." None of the bills made it out of committee, although the movement for reparation was slow to die.[48]

When direct relief was not forthcoming, in the 1950s the struggle shifted to the courts. In 1953 Yuba Consolidated Gold Fields, a major California dredging firm, was preparing a case against the government, but it was never completed.[49] When legal action did come two years later, it was in the form of a suit five hard-rock gold mining companies brought before the Court of Claims in Washington, DC—with 160 additional cases pending the decision. Western dredge concerns, including those in Idaho, eagerly awaited the outcome in the so-called Gold Shutdown Case. When it was handed down, the court found for the companies by a three-to-two margin. A headline in the *New York Times* was clear: "Gold Mines Suit Strikes Pay Dirt," and the editor noted that if the court's order to pay "just compensation" for profits lost during the wartime closing was upheld, settlements could total as much $86 million, in which case the government might have to do some digging in Fort Knox. Further, that figure did not take into account the many future lawsuits that might be brought. Eventually, in 1958 the US Supreme Court overturned the Court of Claims decision and denied compensation, arguing that mining operations had only been suspended and that no property had been confiscated. Thus, the litigation ended but not the bitterness attending Order L-208.[50]

Like most other gold dredges in the country, the Mount Vernon boat lay idle during World War II. Order L-208 was revoked on July 1, 1945, too late for dredge owners to regroup and find skilled crews in the tight labor market.[51] During 1946 Dave Henderson leased his boat to Homa Smith and Paul Bangasser, who then organized Bangasser and Smith, Inc., a Seattle-based venture. They proceeded to dig on leased ground on Crooked River from March 1 until September 7 and recovered 625 ounces of gold and 141 ounces of silver. With gold then valued at $34 per ounce, the take was about $21,250.[52]

Early the next year Henderson sold the dredge to John C. Higgins of H & H Mines, Inc. (Higgins and Hinsdale), a Portland firm that had been digging on the Crooked River since 1941. Like the Mount Vernon Company, Higgins's boat was a two-foot all-steel machine built by Washington Iron Works. It proved to be the leading producer in the district in its initial year, and its next season was a repeat performance. Digging from January 1 to June 20, before the federal ban in 1942, it recovered 2,859 ounces of gold and 560 ounces of silver, a promising output indeed.[53]

Several writers have mentioned a ghost Yuba dredge that operated briefly on the Crooked River, but details of exact time, location, and ownership are lacking.[54] According to the *Minerals Yearbook*, the dredge was a three-and-a-half-foot electric Yuba that belonged to H & H Mines, Inc., which dismantled it in 1942 and moved it to company property on Henderson Creek, Montana, where it continued to dig both gold and strategic scheelite throughout the war and until the end of 1949.[55]

After a slow recovery, the H & H firm resumed dredging on the Crooked River in 1947. The company quickly dismantled the Mount Vernon dredge and moved it to Red Horse Creek, a tributary of the Red River. There it was rebuilt at a cost of $35,000 and set to work on August 15 on ground acquired from Schuyler Simmons. H & H Mines continued to operate on Red Horse Creek throughout 1948 and briefly, for about a month, from May to June 2, 1949, when the dredge was shut down. While working in this area, the firm maintained a small group of quarters familiarly known as Dog Patch, just west of the main street in Elk City, where workers and their families lived.[56]

The boat remained idle until December 1949, when it was dismantled and moved to Buffalo Gulch where H & H Mines started it up again in April 1950, running three shifts a day. How long operation continued is not recorded, but the dredge was idle again in 1951. Finally, in 1952 H & H Mines, Inc., sold the dredge to the Clearwater Dredging Company. At about the same time, H & H Mines disposed of its other holdings, which included ten buildings, and ended its Idaho activity.[57] The Clearwater Dredging Company promptly moved the dredge back to Crooked River, where it ran full-time round-the-clock under the general supervision of William Noon, standing in for Vern Finch, one of the Spokane co-owners. In an abbreviated season between mid-September and the end of December, the boat yielded 840 ounces of gold and 161 ounces of silver.[58] Thus, the Mount Vernon dredge began an auspicious run of several years under the Clearwater banner.

In 1953, when it was the largest producer in the Orogrande district, with a recovery of 3,683 ounces of gold and 672 ounces of silver, Finch's brother,

Claire, was hired as dredgemaster. The next year the boat's output of gold was the highest in Idaho County, its gross recovery valued at more than $109,197; in 1955 its take dropped slightly but was still an impressive $105,000, despite a three-month shutdown because of cold weather. In 1956 it continued on the Crooked River from April 12 to December 4 and was ranked second in the state in gold production. The machine dug only three months in 1957, and its output dropped sharply. At about this time, it was reported that the dredge had been sold to the H & M Company, which supposedly operated on a limited basis.[59]

Precisely why the Clearwater Dredging Company closed down is not clear. The firm had troubles with the State Land Board after the enactment of the first of the new laws, weak though they were, calling for reclamation of dredged ground, and its dredge mining permit had been canceled in 1953, though legal action restored it the following year.[60]

No production was reported for the Mount Vernon dredge in 1958 or 1959, and the date on which it dug its last yard under the auspices of the H & M Company is uncertain. Several possibilities may have prevented the new group from achieving sustained operation. Did the pay streak on Crooked River run out? Was it because of the legislation for reclaiming dredged land? Was it the company's inability to obtain the rights to unworked ground upstream of the dredge? Whatever the reason, the working life of the Mount Vernon dredge ended in the late 1950s. During more than twenty years of digging under different owners, the machine had produced more that 17,000 ounces of gold and silver, with a value greater than $489,000.[61] Abandoned in a quiet pool on the river, the tired old boat was ravaged by rust, decay, and vandals for nearly three decades until 1987, when officials of the Nez Perce National Forest passed the ancient relic along to the West Yellowstone River Gold Museum for permanent display in Nevada City, Montana.[62]

In February 1948 the Warren Dredging Company closed down its four-foot bucket dredge in Lemhi County. During the spring the Warren Dredging Company dismantled it and moved it to the Elk City district, where it was started up on the American River that season between July 6 and November 3, while H & H. Mines ran its two-foot boat on Red Horse Creek from January 1 to December 9. Between them, the two bucket-line machines recovered 3,752 ounces of gold and 650 ounces of silver. During the same year two draglines, one of the Tyee Mining Company on the American River and the other of Hawk Placer in Deadwood Gulch, took out 502 ounces of gold and 88 ounces of silver.[63]

The short initial run of the Warren Dredging Company's Marion dredge was partially a result of equipment failure on the boat and partly a result of unusually bad weather. When a line shaft broke in early November, delay in getting a replacement meant considerable lost time. In January 1949 temperatures of 50°F below zero halted the work of all three shifts for several more idle weeks. By 1950 the Warren dredge was the leading producer in the district, but the following year it finished its ground on the Wild Rose group and was reassigned to the monazite placers of the Cascade Valley.[64]

Notes

1. Waldemar Lindgren, "A Geological Reconnaissance across the Bitterroot Range and Clearwater Mountains in Montana and Idaho," US Geological Survey Professional Paper no. 27 (Washington, DC: Government Printing Office, 1904), 91; Arthur L. Flagg, "The Elk City Mining District, Idaho County, Idaho," *Transactions of the American Institute of Mining Engineers* 45 (New York: American Institute of Mining Engineers, 1914), 113, 116; Stephen R. Capps, "Faulting in Western Idaho, and Its Relation to the High Placer Deposits," Idaho Bureau of Mines and Geology Pamphlet no. 56 (Moscow: University of Idaho, June 1941), 15.

2. William S. Shiach, S. William, Harry B. Averill, and John M. Henderson, *Illustrated History of North Idaho, Embracing Nez Perces, Idaho, Latah, Kootenai, and Shoshone Counties* (Chicago: Western Historical Publishing, 1903), 445; Mary Alfreda Elsensohn, *Pioneer Days in Idaho County* (Caldwell, ID: Caxton, 1947), 1:177, 194; *Idaho Daily Statesman* (Boise), October 6, November 8, December 14, 1896; May 7 and 15, 1897.

3. *Idaho Daily Statesman* (Boise), February 28, 1899.

4. Flagg, "Elk City Mining District," 113–14.

5. *Idaho County Free Press* (Grangeville), February 24, 1899, cited in Elsensohn, *Pioneer Days in Idaho County*, 2:34.

6. Elsensohn, *Pioneer Days in Idaho County*, 1:177.

7. *Idaho Daily Statesman* (Boise), December 29, 1900.

8. Ibid., March 20, 1901.

9. Lindgren, "Geological Reconnaissance across the Bitterroot Range," 91, 93.

10. Elsensohn, *Pioneer Days in Idaho County*, 2:32.

11. *Idaho Daily Statesman* (Boise), April 4, 1900; July 27, 1909; *Davis County Clipper* (Bountiful, UT), October 18, 1901.

12. *Engineering and Mining Journal* 29 (November 7, 1908): 719; *Mining and Scientific Press* 98 (January 23, 1909): 135.

13. *Salt Lake Mining Review* 10 (January 15, 1909): 12, 15; *Salt Lake Mining Review* 11 (July 15, 1909): 3; *Salt Lake Mining Review* 11 (August 30, 1909): 12; *Salt Lake Mining Review* 11 (November 30, 1909): 17; *Eleventh Annual Report of the Mining Industry of Idaho for the Year 1909* (State Mine Inspector's Report), 62; *Mining and Scientific Press* 98 (April 3, 1909): 469; *Mining and Engineering World* 29 (October 31, November 7, 1908): 681, 719; *Mining American* 58 (October 15, 1908): 312. See also Elk City Dredging Company Stock Certificate, 1909, at https://www.fold3.com/document/232535613/.

14. Flagg, "Elk City Mining District," 113–14.

15. *Salt Lake Mining Review* 11 (July 30, 1909): 16.

16. Ibid. (November 15, 1909): 17; *Mineral Resources, 1909* (US Geological Survey), 348.

17. *Mining and Scientific Press* 98 (March 27, 1909): 435.

18. *Salt Lake Mining Review* 11 (November 30, 1909): 16–17; *Mining and Scientific Press* 100 (January 1, 1910): 69. *Mineral Resources, 1909*, 348, says that neither of the dredges was able to work before winter struck.

19. *Mining and Scientific Press* 100 (January 1, 1910): 69; *Sunset* 24 (San Francisco: May 1910), part 2, advertising section, no page number.

20. *Mineral Resources, 1910* (US Geological Survey), 460; *Mineral Resources, 1911* (US Geological Survey), 593; *Mineral Resources, 1912* (US Geological Survey), 726; *Mineral Resources, 1913* (US Geological Survey), 775; *Mineral Resources, 1914* (US Geological Survey), 622; *Mineral Resources, 1915* (US Geological Survey), 545; *Mining and Engineering World* 44 (January 1, 1916): 35.

21. Francis C. Nicholas, "An Unpopular Land Office Decision in Idaho," *Mining World* 22 (February 12, 1910): 381–82; *Mining and Scientific Press* 104 (June 29, 1912): 904.

22. *Idaho Daily Statesman* (Boise), May 1, 1914.

23. *Mineral Resources, 1915*, 545; *Salt Lake Mining Review* 17 (March 30, 1916): 26; *Mining and Engineering World* 44 (January 1, 1916): 45.

24. *Spokesman-Review*, October 3, 1919, clipping, Kimble Database, WSU.

25. Stewart Campbell, *Twenty-Third Annual Report of the Mining Industry of Idaho for the Year 1921* (State Mine Inspector's Report), 6.

26. Ibid.

27. Quoted in *Salt Lake Mining Review* 26 (March 30, 1925): 7.

28. *Idaho Daily Statesman* (Boise), September 12, 1912; *Salt Lake Mining Review* 26 (March 30, 1925).

29. *Spokesman Review*, October 10, 22, 1929, clipping, Kimble Database, WSU.

30. Sarah Elizabeth Purdy, "Analysis of Dredge Tailing Pile Patterns: Applications for Historical Archaeological Research," MA thesis, Oregon State University, Corvallis, June 7, 2007, 40.

31. *Spokane Chronicle*, October 6, 1937, clipping, Kimble Database, WSU.

32. Ibid., October 6, November 9, 1937, clippings, Kimble Database, WSU.

33. Other businesses included in the consolidation were the Mineral Products Company, Idaho Exploration, Inc., the Gold Production Company, and the Metals Syndicate—all Idaho firms that controlled mining properties in Idaho, Oregon, or Montana. Ibid., May 4, 1938, clipping, Kimble database, WSU.

34. Purdy, "Analysis of Dredge Tailing Pile Patterns," 56–58; *Minerals Yearbook, 1938* (US Geological Survey), 352; *Minerals Yearbook, 1939* (US Geological Survey), 328; *Minerals Yearbook, 1941* (US Geological Survey), 350; *Minerals Yearbook, 1942* (US Geological Survey), 380; Stephanie Van Buskirk, "Historic Context for the Upper Newsome Creek Dredge Mining Site, Idaho County, Idaho," 21–22, in "Newsome Mining Report," September 2005, at ftp://ftp.nezperce.org/.../Newsome%20EIS/Newsome%20Mining%20.

35. *Minerals Yearbook, 1941*, 350; *Minerals Yearbook, 1942*, 380; *Minerals Yearbook, 1946* (US Geological Survey), 1441; *Minerals Yearbook, 1947* (US Geological Survey), 1409; *Minerals Yearbook, 1948* (US Geological Survey), 1502; *Minerals Yearbook, 1949* (US Geological Survey), 1474; *Minerals Yearbook, 1950* (US Geological Survey), 1495; *Minerals Yearbook, 1952* (US Geological Survey), 299; *Minerals Yearbook, 1953* (US Geological Survey), 343; *Spokane Daily Chronicle*, November 30, 1950; F. C. Armstrong and P. L. Weis, "Uranium-Bearing Minerals in Placer Deposits of the Red River Valley, Idaho County, Idaho," *Geological Survey Bulletin* 1046-C (Washington, DC: Government Printing Office, 1957), 25–26, 35.

36. Armstrong and Weis, "Uranium-Bearing Minerals in Placer Deposits," 51, 54; *Idaho County Free Press* (Grangeville), June 30, 1976; Rolland R. Reid, "Placer Deposits of the Elk City Region," Idaho Bureau of Mines and Geology Pamphlet no. 121 (Moscow, ID: Idaho Bureau of Mines and Geology, July 1960), 7.

37. Van Buskirk, "Historic Context," 25–26, 30.

38. *Minerals Yearbook, 1937* (US Geological Survey), 323–24; *Kellogg (Idaho) Evening News*, clipping, Kimble Database, WSU.

39. *Minerals Yearbook, 1932–33* (US Geological Survey), 107; State of Oregon, *List of Mines in Oregon* (Portland, OR: Department of Geology and Mineral Industries, March 1939), 13; John C. Reed, "Geology and Ore Deposits of the Florence Mining District, Idaho County, Idaho," Idaho Bureau of Mines and Geology Pamphlet 46, Moscow, ID, 43–44; Oregon Department of Geology and Mineral Industries, *The Ore-Bin* (Portland, OR: Ore-Bin, February 1941), 3:13; *Lewiston (Idaho) Morning Tribune*, September 9, 1936; *Spokesman-Review*, September 10, 1937, clipping, Kimble Database, WSU.

40. *Minerals Yearbook, 1937*, 323; *Minerals Yearbook, 1938*, 352; *Minerals Yearbook, 1939*, 329; *Minerals Yearbook, 1940*, 344; Van Buskirk, "Historic Context," 33–35.

41. Purdy, "Analysis of Dredge Tailing Pile Patterns," 51, 59; *Minerals Yearbook, 1935* (US Geological Survey), 207. In 1935 Henderson was also president of the Orogrande Company, which owned a massive diesel-powered shovel built by Washington Iron Works at a cost of $100,000 and billed as "The Most Modern for Placer Mining," *Spokane Chronicle*, July 16, 1935.

42. H. S. Lorain and O. H. Metzger, "Reconnaissance of Placer Mining Districts in Idaho County, Idaho," *US Bureau of Mines Information Circular* 7023 (June 1938): 43–45; Phoebe Siddall, *South Fork Clearwater River Habitat Enhancement Nez Perce National Forest* (Portland: US Department of Energy, April 1992), 5; Sharon A. Murray, "The Mount Vernon Dredge," typescript MSS, 1985 or 1986, 1–2, copy, on file at Nez Perce National Forest Supervisor's Office, Grangeville, ID; *Minerals Yearbook, 1936* (US Geological Survey), 39.

43. Murray, "Mount Vernon Dredge," 2–3; *Minerals Yearbook, 1938* (US Geological Survey), 352–53; *Minerals Yearbook, 1939*, 329; *Minerals Yearbook, 1941*, 350; *Minerals Yearbook, 1942*, 380.

44. Office of War Information and War Production Release on Gold Mine Shut-Down Order (October 8, 1942), copy, Alaska Gold Company MSS, Fairbanks, AK; *New York Times*, October 8 and 9, 1942.

45. *Engineering and Mining Journal* 144 (January 1943): 92; Elsensohn, *Pioneer Days in Idaho County*, 1:103–4.

46. Maryclaire McCauley, *The Closing of the Gold Mines, August 1941 to March 1944* (Washington, DC: War Production Board Special Study 9, first issued June 1, 1944; reissued April 5, 1946), 12–16, 33–34.

47. See advertisement for Yuba Manufacturing Company, *Engineering and Mining Journal* 146 (December 1945): 18; "Rosie the Riveter: United States History," n.d., http://www.u-s-history.com/pages/h1656.html; Harold F. Williamson and Kenneth H. Myers II, *Designed for Digging: The First 75 Years of Bucyrus-Erie Company* (Evanston, IL: Northwestern University Press, 1955), 270–71, 274, 276.

48. *Engineering and Mining Journal* 148 (November 1942): 43; *Congressional Record* (June 11, 1945), 79th Cong., 1st sess. (1945), 5919–20; *Congressional Record* (October 16 and 18, 1945), 79th Cong., 1st sess. (1945), 9726, 9761; *Congressional Record* (February 26, 1946), 79th Cong., 2nd sess. (1946), 1594–95; *Congressional Record* (January 14, 1947), 80th Cong., 1st sess. (1947), 329; *Engineering and Mining Journal* 147 (March 1946): 102; *Engineering and Mining Journal* 150 (April 1949): 128; *New York Herald Tribune*.

49. Allen R. Molten to S. R. Pound, San Francisco, March 13, 1953, copy; Molten, Proposed Complaint in the L-208 Case, April 2, 1953, draft, both in Alaska Gold Company MSS, Rasmuson Library, University of Alaska, Fairbanks.

50. *United States v. Central Eureka Mining Company et al.*, 367 *U.S. Reports* (1958), 155–84; *Engineering and Mining Journal* 147 (March 1956): 106; *Engineering and Mining Journal* 149 (July 1958): 126, 128; *New York Times*, February 21, 1956; June 17, 1958.

51. Clark A. Spence, *The Northern Gold Fleet* (Urbana: University of Illinois Press, 1996), 119.

52. Murray, "Mount Vernon Dredge," 3; *Minerals Yearbook, 1946*, 1441.

53. Murray, "Mount Vernon Dredge," 3–4; Purdy, "Analysis of Dredge Tailing Pile Patterns," 59; *Minerals Yearbook, 1941*, 350; *Minerals Yearbook, 1942*, 380.

54. Purdy, "Analysis of Dredge Tailing Pile Patterns," 51; Siddall, *South Fork Clearwater River Habitat Enhancement*, 5.

55. *Minerals Yearbook, 1943* (US Geological Survey), 399.

56. Murray, "Mount Vernon Dredge," 3–4; *Minerals Yearbook, 1947*, 1409; *Minerals Yearbook, 1948*, 1502; *Minerals Yearbook, 1949*, 1474.

57. *Engineering and Mining Journal* 151 (February 1950): 154; Murray, "Mount Vernon Dredge," 4; *Minerals Yearbook, 1947*, 1409; *Minerals Yearbook, 1948*, 1502; *Minerals Yearbook, 1949*, 1474; *Minerals Yearbook, 1950*, 1481, 1495; *Minerals Yearbook, 1951* (US Geological Survey), 1493; Purdy, "Analysis of Dredge Tailing Pile Patterns," 59.

58. Murray, "Mount Vernon Dredge," 4; *Minerals Yearbook Area Reports, 1952* (US Geological Survey), 299.

59. Murray, "Mount Vernon Dredge," 5–6; *Minerals Yearbook Area Reports, 1953* (US Geological Survey), 342–43; *Minerals Yearbook Area Reports, 1955* (US Geological Survey), 338, 357; *Minerals Yearbook Area Reports, 1956* (US Geological Survey), 378; *Minerals Yearbook Area Reports, 1957* (US Geological Survey), 363, 376.

60. Murray, "Mount Vernon Dredge," 6; *Minerals Yearbook Area Reports, 1957*, 376.

61. Murray, "Mount Vernon Dredge," 6.

62. Despite a long article in the Elk City newspaper in early 1986, based completely on Sharon Murray's detailed report, the editor's plea for Idaho residents to reclaim the dredge as a historic artifact went unheeded. Forest officials had claimed possession on grounds of abandonment, and once local citizens had shown their lack of interest, the officials were happy to see the machine transferred to Montana. For the article, see clipping from the *Idaho County Free Press*, January 22, 1986, copy, in Nez Perce National Forest Supervisor's Office, Grangeville, ID.

63. *Minerals Yearbook, 1948*, 1502.

64. *Engineering and Mining Journal* 150 (March 1949): 115; *Engineering and Mining Journal* (May 1949): 134; *Minerals Yearbook, 1950*, 1495; *Minerals Yearbook, 1951*, 1479–80.

7

Stanley Basin and Yankee Fork

The Stanley Basin is a wide-open valley in high mountain country. Bordered by three mountain ranges with forty peaks over 10,000 feet above sea level and 300 high lakes, its scenic beauty is unsurpassed, a young Ed Burroughs noted in 1903.[1] Streams flowing from the Sawtooth Mountains merge in the valley to form the Salmon River, the largest tributary of the Snake. Placer gold was first discovered there in the early1860s, and quartz mining soon followed, but the beginnings were modest and developed slowly. Marsh Creek, Valley Creek, Warm Springs Creek, and Robinson bar showed real promise, but miners were hampered by short working seasons as a result of harsh winters and sometimes by the lack of water.

In the waning years of the nineteenth century, both gold placer and quartz lodes attracted more attention, especially when the new dredge technology began to catch on. Promoters with a little capital bonded placer ground along the streams and in the flats and hoped to sell at a profit to companies organized to dredge the property. In late 1897 Sid Roberts of Boise, Joe Montgomery of Hailey, and Carl Lane of Challis combined to bond a total

DOI: 10.5876/9781607324751.c007

of 365 acres on Stanley Creek from four separate owners; they also acquired valuable water rights to work the higher bars in the basin. Lane, the sheriff of Custer County, had an interest in the bond as well as the property. The ground was flat, and Roberts considered it "probably the finest dredging proposition in the northwest." There was paydirt from "grass roots to the bedrock, some 13 feet," Roberts said, and his own test of the gravel showed a return of forty-seven cents per yard, while the tests of others ran even higher. One appraisal assigned a value of more than $5 million for the 365 acres.[2] These were promoters' estimates, of course, which the wary would take with a proverbial grain of salt.

Subsequently, the three holders of the bonded ground had shown the property to two Montanans—a potential investor named Monahan and S. E. Johnson, representing the Harper Dredge Company in Bannack.[3] There may have been some confusion about the name of the firm. Samuel S. Harper was no doubt the man for whom the company was named. Harper was also the person largely responsible for designing the first successful bucket dredge in the United States, the *Fielding L. Graves,* which made its debut at Bannack in May 1895.[4] He took the lead in organizing the Gold Dredge Company to run the boat, but there is no evidence that Harper Dredge Company, as such, was digging gold. Perhaps it existed to build dredges or was a new business firm seeking a place to locate in Idaho. Harper did have some developing dredging interests in the Boise Basin near Placerville.[5] There is no indication that the Montana visitors from Bannack proceeded any farther with negotiations in the Stanley Basin.

In late 1898 a Boise newspaper carried a brief notice that a dredge would be built in the basin and be in operation the next spring. This would be the work of the Stanley Basin Dredging Company, at first called the Kansas City Dredging Company, then sometimes referred to variously as the Sawtooth Dredging Company or the Stanley Dredging Company by newspapers. The Stanley Basin Dredging Company may have had some Kansas capital, but it was registered in Perry, Oklahoma, and backed mainly by capitalists from that territory. Its president was S. A. Wickard and its secretary was George A. Masters, both involved in the business of buying, storing, and selling broom corn. The dredge site would be on the upper part of Stanley Meadows. Material had already been brought to the ground to build the boat, and great accomplishments were claimed for this new type of machine. The editor extolled the virtues of a similar rig, reputedly capable of saving up to 98 percent of the gold, already in place on Moore Creek in adjoining Lemhi County. "If this machine will do one-half the work claimed for it[, it] will make a revolution in placer mining in Stanley," he said, a fairly standard platitude.[6] Construction was not completed on schedule, but the company promised a second boat, once its sawmill could turn out the lumber. The new dredge was belatedly launched in mid-August 1899, even before the last of its machinery arrived from Kansas City. Superintendent Ed Lynn presided, with the honor of the christening conferred upon eleven-year-old Eva Roberts, daughter of Sid Roberts, the Boise promoter. The *Pearl L*, as the dredge was called, was expected to begin digging a month later or by October 1 at the latest, and a crew of thirty men was at work to meet the deadline.[7]

There is little evidence that other dredges were actually completed at this time. The dredge was not identified as such, but it was no doubt the *Pearl L* that in two months of the previous year had raised over $800 a day, with six men working and the engine burning ten cords of wood per diem. It had to be the same boat Sid Roberts was referring to when he mentioned the Sawtooth company and its work in 1900. The corporate backers were Oklahomans, and his description fit the history of the *Pearl L*. The boat ran for about two months "in more or less of an irregular manner" but cleaned up about $6,500 from ground returning sixty cents per yard. The gravel was richer than that, according to Roberts, but the ladder was too short to reach bedrock, and other defects in the machinery, plus poor handling, left much to be desired. The dredge, with a capacity of 2,500 yards a day, had handled only about an eighth of that amount in a season that was "experimental to a degree." All of these shortcomings were to be remedied before the next year, said Roberts.[8]

But the dredge was in worse shape than Roberts realized. In late June 1901 the *Deseret News* reported that repairs begun earlier had been completed and that the machine would be ready to go in a few days. Apparently, it dug until mid-September 1899, when operations were suspended for the season. A spokesman for the company explained that a new dredge was needed, that the one built in Kansas City was too weak. Not only was the bucket ladder too short to reach bedrock, it was also much too light. Under the strain of full power it broke, and replacement lengths did likewise almost as quickly as they were put in. Thus, the dredge could not handle more than 300 yards or

400 yards a day instead of its 1,500-yard capacity. Even so, cleanups averaged $150 per day, yield was 51 cents a yard for the truncated season, and major expenses were wages for four men and the cost of four cords of wood daily delivered to the boat.[9] This seems to have been a death knell for the *Pearl L* and its corporate sponsors.

Meanwhile, others were investing in the Stanley Basin placers. Sid Roberts had an interest in property on Stanley Creek adjoining that of the Kansas City Dredging Company, where he and George Smith apparently had a sluicing operation that yielded $9,000 in 1900 before they ran out of water. In their minds, the ideal solution for the water shortage would be to find capitalists willing to finance a ditch fifteen miles from Valley Creek and to bring in dredges. Ground also adjacent to the Kansas City outfit was owned jointly by J. M. Ballentine and E. J. Watters. Ballentine owned two farms near Boise and had served two terms in the Idaho legislature. Watters, originally from Little Rock, promoted mines and was heavily into Wyoming oil. In September 1899 the two partners were at the scene overseeing six men at work drilling prospect holes in hopes of proving that the ground was dredge-able when Ballentine died of a heart attack. At the same time, the Idaho City Dredge Company had six or seven men testing property on Valley Creek, above the mouth of Stanley Creek. At the head of Warm Springs Creek, a tributary of the Salmon River, a Challis group located the entire Fisher Basin, and a dredge company was expected to take over in the summer of 1900.[10]

But apart from the *Pearl L*, in none of these instances did drill holes lead to actual dredges until 1903, when George Burroughs created the Stanley Basin Mining Company and moved the updated *Yale* from the Snake up to Stanley Creek near the site of the original discoveries of 1864. Here, the free gold was not the fine flakes of the Snake but larger, "of the mustard seed variety," as someone described it. Although the press extolled the venture and noted that the rig "has proven to be a great mechanical success and a money-maker," the *Yale* dug only a part of a season, then closed down.[11] Burroughs gave no reasons for the closure.

Beginning two years after Burroughs's departure, *Mineral Resources of the United States* for 1905 and 1906 indicated that the main work in the basin was in Stanley Creek and Joe's Gulch, adjoining areas, and that the Womack dredge on the property of the Stanley Basin Dredging Company was the only rig of the kind running. The other major producers were sluicing or hydraulicking.[12] A year later, in 1907, the same publication repeated that same statement for the same ground, except that now the Saw Tooth Placer Mining Company was operating the Womack boat. The firm was incorporated in Kansas City, Missouri, with capital of $300,000 and a board of directors that included H. J. Prange, W. M. Rynerson, W. A. Moses, B. E. Moses, and L. E. Moses.[13] No further explanation was given about the machine or the people involved.

The undated prospectus of the Idan-ha Gold Dredge Company may have been issued as early as 1905. It laid out in detail its specific plans for the Stanley Basin. The company had nominal capital of $800,000 of $1 par value and had bonded 540 acres along both sides of Stanley Creek for more than two miles. Its mission was to work that ground with one or more "Standard Gold Saving Dredges." Corporate officers included O. Robinson of Richmond, Virginia, president; D. C. Young Jr., vice president; W. L. Grose, secretary; and H. C. Willis, treasurer and manager. All except Robinson were listed as from Boise, although Willis, the driving force, was a transplanted New Yorker, an electrical engineer by training.[14]

The prospectus was typical of the genre of its day. The company's property contained about 10 million cubic yards of gravel, each containing "high values in *Gold* . . . in one of the *richest* and most *extensive* placer deposits in the United States." The ground had been carefully tested and showed values averaging 50 cents a yard, with some running as high as $15 per yard. Next came the mathematics. For starters, assuming capacity to handle 1,500–1,600 yards a day, one dredge could make $20,250 in a month from free gold, which was not the fine flakes of the Snake. Add to that an additional $77,625 a month from black sands for a total of $97,875 a month. Thus, for six and a half months of operation, a whopping $636,187.50 would available for dividends—"not an exaggeration, but *plain, cold, hard facts.*" Shares were available for only 50 cents apiece. The prospectus was topped off with four pages of testimonials from local experts, including one from manager-to-be Henry Willis.[15]

Still, it was a legitimate enterprise, not one developed merely to profit from stock. The new corporation found backing, apparently from Wisconsin capitalists, and proceeded with its business. In the summer of 1908 there were two large dredges in the district owned by "an eastern company," probably the Idan-ha group, in addition to one belonging to Matt Womack of Boise. Willis brought in a small, new type of dredge to supplement the standard bucket-line boat, one that did not float in a hull, and he was agreeably surprised at its performance. After a dozen or so days of experimenting, the new machine was running twenty hours a day, handling three cubic yards per minute. A dry-land dredge, which moved on a track and was fitted with a drag scoop, the machine had serious problems, however. The gravel rested on clay bedrock, and the scoop dumped chunks of clay so large that they would not disintegrate, with the result that most of the gold values were carried into the tailings pile. In the end, the rig was considered a failure and had to be abandoned.[16] The gold production of the Stanley Basin that season was valued at $4,138, nearly all from placers, compared with $10,296 the previous year. The drop was attributed to the idleness of the Saw Tooth Placer Mining Company dredge. The Idan-ha Company was succeeded by the Stanley Dredge Company, which produced no gold in 1909 because its equipment proved too antiquated and inefficient; its replacement, the Willis Company, planned to build a new dredge.[17]

In 1910 the Willis Placer Company, as it was called, encountered a layer of talcose gravel containing rusty gold, which forced it to add disintegrating devices to its machine; during the following season its dredge operated for only a few days. Over the next two years, much time was devoted to adding improvements for saving gold and platinum from the recalcitrant soil and making intermittent test runs. Experiments showed that the black sand residue also contained high values and that cinnabar sands—mercury sulphides—were present in paying quantities if the modifications proved effective. Willis thought they would. He believed that a simple, practical, and economical method of recovery had been worked out and that the 1913 season would be very profitable. The boat was expected to start up in mid-June and to employ twenty-one men. During 1914 it did not operate at all, however. The following year it closed down.[18]

For the next five years Henry Willis seems to have vanished from the mining scene—or at least his presence was obscured. In 1921 he reappeared as manager of the Sawtooth Dredging Company, a concern that owned substantial acreage of ground and a "partly completed steam driven dredge." Willis was running a Bucyrus on Stanley Creek across the divide from Joe's Gulch and reputedly recovering values of between thirty cents and a dollar a yard. Its 2,000-yard-capacity boat was fitted not only with the normal gold-saving equipment but also with special tables and screens for capturing the black sands, which were reported to contain both gold and platinum. The next year, Willis was no longer in charge of the dredge but had moved to Berkeley, California. He still served as secretary of the company, in which he no doubt continued to have a vested interest. H. U. Davis of Twin Falls was president and R. Davis was manager.[19]

In 1922 the Sawtooth Dredging Company owned five-and-a-quarter miles—"the entire length of Stanley Creek," including Buckley bar, from whence $700,000 had been taken since its discovery in the early 1860s. The Sawtooth crews were scheduled to start up their small boat in mid-May 1922. Whether it dug much that season is not recorded, but it ran only a short time in 1923.[20]

In early August 1928 the Spokane *Spokesman-Review* announced a major mining deal. The old Sawtooth Dredging Company property in Stanley Basin had been sold by F. N. Johnson to C. H. Shaw of San Francisco for $150,000. Involved was a block of 1,200 acres formerly belonging to the Spreckels interests. Johnson had purchased the ground the previous October for $45,000 after the Sawtooth Dredging Company had gone into receivership. Shaw's engineers had spent four months testing the ground and reported favorably that it would yield 40 cents per yard. Shaw intended to erect an electric power plant and install two electric dredges on the property as soon as possible, he said.[21] It never happened.

Meanwhile, several earlier attempts had been made to induce capital for dredging, without much practical success. It was reported in early 1909 that the Greyhound Mining Company at Seafoam, in the rugged area high above the Middle Fork of the Salmon River, had plans for a dredging operation in some undisclosed part of Stanley Basin—something that also never materialized.[22]

At about the same time, the *Statesman* ran a lengthy piece about an effort by J. H. Andreas to negotiate a deal to sell a rich stretch of ground near Robinson bar on the Salmon in the Stanley Basin. Robinson bar was well-known and had already produced at least $1 million worth of gold. The property was owned mainly by Boise capitalists, who were asking $15,000 for it; the potential buyers were unnamed and from the East. Andreas noted that "mining deals" were rare and frowned upon in the basin. It was difficult to interest local people in placer investments; in recent years the focus had been on irrigated properties. Now, the pendulum was swinging. Improvements in dredge technology were increasing the demand for large blocks of suitable land, according to Andreas.[23] Whether a deal was ever struck is uncertain. In August 1913 mining expert J. Asher Cowan had just returned from Stanley Basin where he had examined a large placer proposition, complete with a dredge that had a 1,000-yard capacity per day. The boat made a three-day run with satisfactory yields, showing values of from 50 cents to $1.10 a yard and saving a high percentage of the gold.[24] Precisely where the dredge was located, who owned it and the ground, and who the possible purchaser was were never stated.

The Idaho Dredging Company was organized in Spokane in early October 1928 as a holding corporation "to protect and develop" 750 acres of placer ground near Stanley on the headwaters of the Salmon River, according to President Charles LeMarinel. Incorporated with nominal capital of $250,000, the firm's property was considered extremely valuable by its officers. According to LeMarinel, recent assays indicated gold values of around seventy cents per cubic yard, and the ground "is almost level and ideally situated for dredge operation." He hinted that the company might be willing to lease the ground.[25] In the absence of evidence of any working sale or operation, this was probably another pipe dream.

Between 1933 and 1938, C. H. Lord and associates controlled 120 claims of the old Willis ground along Stanley Creek. The group ran a one-cubic-yard bucket dragline, two bulldozers, and a large trammel washing plant, with an overhead cableway to dispose of tailings—all with disastrous economic results. Lord is reported to have spent $350,000 and recovered only $23,000 worth of gold. Like Willis, he was never able to solve the "robber" clay problem and separate the precious metal.[26]

In the summer of 1941 the Pilot Gold Dredging Company of Prairie City, Oregon, moved its boat to Valley Creek, about fifteen miles from Stanley.[27] There is no indication in the US Geological Survey's official documents that the machine was ever active, although it is possible that this was the dragline and floating dredge that ran on the Elk Creek Placer during at least part of the 1941 season.[28]

In the Stanley Basin, on ground owned by Jason Vaught, Jordan Placer Company, Inc., operated a dragline on a mile and a quarter of lower Jordan Creek between 1948 and 1950. The rig had a one-cubic-yard bucket and a seventy-ton washing plant mounted on caterpillar tracks. During two five-month periods in 1948 and 1949, a total of $233,000 in gold was recovered. How much was taken out in 1950 is not recorded.[29]

Slightly northeast of Stanley Basin, placer mining was on the eve of revival in the Yankee Fork, another branch of the upper Salmon. Successful gold dredging was slow to catch on in the area, but eventually, by the late 1930s, one company was able to accomplish a sustained run. A start had been made much earlier, especially from Jordan Creek south down the river several miles below the town of Bonanza, where a number of old-timers relocated a large amount of abandoned ground in the hope of catching the eye of dredge promoters. The upshot was that in the autumn of 1899 the Souther Dredge Company of Boston, operating successfully in the Boise Basin, sent J. N. Wallace to put down test holes for two miles down Jordan Creek. The company was so impressed with Wallace's findings that it either bonded or filed on as much vacant land as possible near Bonanza.[30] When a Hailey group took an option on some available ground near the same town in 1901 and sought funds to prospect and examine it, local mining men cheered and saw dredging as the way of the future.[31] It was indeed some distance in the future.

In 1926 the New York Engineering Company obtained options on the ground and sent engineer D. H. Ferry, who was employed by the Guggenheims running a dredge near Murray, farther north in the state. Ferry put down twenty-seven drill holes and three shafts and reported an average of 17 cents a yard for 11 million acres. The $150,000 required to proceed would be returned in eighteen months, he said, but despite Ferry's strong recommendation, the New York Engineering Company took no action.[32] When the president of the firm was queried about the property in 1933, he indicated that it was not suitable, that several rocky reefs cut across the ground, making dredging a precarious venture. Moreover, Ferry's estimates of values were much too high.[33]

In May 1934 a Spokane newspaper reported that a purchase was being negotiated for dredging on property test drilled many years earlier on the Yankee Fork, according to mining engineer F. Cushing Moore.[34] Later, either the same year or in 1935, state senator Ralph E. Whitten was instrumental in consolidating the earlier options into a solid block and forming the Yankee Fork Placer Mining Company to work the ground, which was no doubt the same stretch that Ferry had evaluated. The new firm had a small dredge shipped from the Yuba Manufacturing Company in California to Idaho, but apparently no further examination had been made. Workers proceeded to build a bunkhouse and a cookhouse, but when they began to dig the pit for installation of the boat, they found so many large rocks that the engineer in charge decided the ground was impossible for dredging. Thus, the company found itself with a new modern dredge and no ground.[35] Charles Janin, one of the best-known experts in the field, was critical: "The Yuba crowd are [*sic*] pretty snooty," he proclaimed. He faulted that group for the fiasco: "The dredge was a three-foot one and the ground has as I put it six-foot boulders—this was a rotten business[;] everybody blames some one else."[36] After flirting with the idea of moving the boat to the John Day River in Oregon, the unfortunate syndicate was able to unload its three-footer, at a considerable loss, to a new company being formed near Wallace, California.[37]

The increased price of gold in 1935, combined with the lower wages and material costs that prevailed during the Depression, caused gold mining to become attractive again. Old mines reopened and currently operating mines expanded. Owners of placer claims on the Yankee Fork renewed their efforts to sell their property and to apply more expensive equipment, namely bucket dredges, which had long since proved their worthiness. Other companies soon showed an interest. In 1937, realizing that the options of the Yankee Fork Placer Mining Company were about to expire, the firm of Baumhoff and Fisher contemplated moving one of its dredges from Warren, but after inspecting the ground, it settled on Moose Creek near Salmon.[38] Eventually, a corporate buyer materialized—the Silas-Mason Company, a New York construction firm fresh from helping to build Grand Coulee Dam, with pockets that were correspondingly deep. The firm began testing the ground in the fall of 1938 and resumed the next spring, after a winter break. Gold content ran from $0.093 to $0.689 per yard—certainly a good showing for a potential dredging site and enough inducement for Silas-Mason to organize a subsidiary, the Snake River Mining Company. Ralph E. Whitten was again a major shareholder in the firm.[39]

In the autumn of 1939 the new corporation solicited bids for an appropriate dredge from both the Yuba Manufacturing Company in California and Bucyrus-Erie in Wisconsin. Yuba dropped out of the running when it remembered its earlier debacle of the too-small three-foot machine, and it refused to send an expert to look at the ground. Bucyrus-Erie agreed to design and construct a heavy eight-foot bucket-line dredge capable of handling large boulders and digging thirty-five feet below the water line. Meanwhile, the Snake River Mining Company proceeded to provide modern living facilities for construction workers, dredge crews, and their families. At the same time, Olsen City Steel of Boise constructed the twenty-five steel pontoons on which the dredge would float. The logistics of assembling all the components were challenging but not unique to Yankee Fork; by this late date, American dredge machinery had been hauled laboriously to countless remote wilderness areas the world over—from Alaska to Siberia, Africa, and the jungles of New Guinea. Bucyrus-Erie shipped its equipment by rail to Mackay, 100 miles from the dredge site. From the railhead at Mackay, the Lindburg Truck Line used a specially strengthened vehicle and a four-wheel-drive Ford to haul the heavy machinery the rest of the way over steep, narrow, dirt mountain roads. All of the pontoons and some structural steel from Boise were trucked the entire distance to Yankee Fork.[40] Once all the materiel was on-site, assembly proceeded smoothly, ironically beginning in the pit intended for the small dredge abandoned earlier. A crew of forty-three men completed construction of both the dredge and the camp in four months in late August 1940.[41]

Once finished, the dredge was a far cry from the turn-of-the-century boats that had preceded it in the Stanley Basin and on the tributaries of the Salmon elsewhere. Although not the largest gold dredge of its time, the big machine weighed 988.125 tons and cost $390,244.97, of which $45,259.35 was for labor in transporting and building the rig. With the price of building the camp with modern living quarters plus additional non-dredge equipment, the total outlay came to $428,304.27.[42]

In appearance, the dredge looked like other modern gold dredges, although a writer for the *New York Times* much later described it in tenderfoot terms: "If a battleship had been designed by Rube Goldberg and then beached in a remote gully, it might look something like this contraption."[43] Powered by two Ingersoll-Rand diesel electric units, which produced 450 horsepower at sea level but only 347 horsepower at Yankee Fork's altitude of 6,000 feet, together they consumed 600 gallons of fuel in a three-shift day. Main tanks in several of the pontoons held 18,000 gallons, with special pump apparatus to move diesel oil to day tanks higher up. The ladder hoist, upper tumbler drive, rotary screen trommel, stacker, and other parts of the dredge were powered by separate electric motors provided by the General Electric Company, with sophisticated gearing and controls. All of the equipment was heavy-duty. The seventy-one buckets weighed over a ton each and were held in place by pins five-and-a-half inches in diameter. The bucket ladder was equipped with one of the first automatic high-pressure lubrication systems, although other moving parts continued to be tended by two oilers armed with hand-held grease guns. Another modern innovation was hydraulic controls for the numerous winches that activated the various parts of the machinery. Gone were the heavy clutch and friction brakes that made the winchman's task a strenuous one.[44]

The dredge crew was typical of the industry. The most important member was the dredgemaster—the experienced professional who was in charge, made the decisions, and was on call for all three shifts. Each shift had a winchman, also a key and seasoned employee, who handled the controls of the monster rig. There were two oilers for each shift. The bow oiler was responsible for the two diesel engines and all grease fittings in the front half of the boat, while the stern oiler took care of the rest—especially the trommel, which was prone to clogging, and the stacker, which had a large number of fittings. On the day shift, a cat skinner made sure the path of the dredge was clear of trees and brush and also took care of digging in the boat's "dead men," its buried anchors onshore. A machinist for shop repairs was vital, and often the day shift also included a gold man to pan samples, tweak the sluices, and handle cleanup recovery of the gold.[45]

Although the dredge was state of the art, like all new complex machines it was not without complications in its early days of operation. Equipment needed to be adjusted and fine-tuned. Many of the pipes carrying water pumped through the various cooling systems and vital to wash the gravel leaked heavily. Gold recovery was much less than anticipated, so the company replaced the cleanup man with a California expert who tinkered with the tables, switched to Hungarian riffles, and added quicksilver traps. Values saved soon went up. Large boulders were a problem from the beginning and led to constant repairs of the damage to the dredge. The company blamed faulty design by Bucyrus-Erie, whose spokesmen responded by noting that the manufacturer had recommended special rock-removal gear, but that idea had been vetoed by the president of the Snake River Mining Company, who had assured B-E engineers that the boulders would be removed by other machinery before the dredge commenced digging. Then, too, there were troubles with the diesel fuel provided by the Yale Oil Corporation, a Billings, Montana, outfit. Despite the use of additives recommended by the oil company, after a few months the engines began to corrode, leading to repair bills and eventually a broken contract and a lawsuit.[46]

All the while, after startup in 1940 and 1941, the dredge was digging gold, albeit with uncertainty as war clouds threatened. The United States was grooming itself to enter World War II as the second half of 1941 proceeded, and the mining of placer gold especially had a low-priority rating when it came to spare parts. It took a good deal of lobbying by Ralph E. Whitten and Idaho governor Chase Clark before the company could obtain permission late in the year to obtain the parts it needed to fill its needs. Although the boat continued to operate for almost another year, it was a temporary remission. As of October 15, 1942, the War Production Board closed down all nonessential mining. Thus, the Snake River Mining Company dredge sat idle until the spring of 1946, when operations resumed.[47]

Working upstream, despite higher costs for labor, fuel, and parts, the boat dug through the rich ground of Jerrys Creek meadow but hit poor gravel in 1947, which gave company officials pause. They could either buy claims further up the stream or consider selling the dredge. By mid-year the decision had been made—sell out and focus on tungsten property in New Mexico. The boat was moved to a pond near the juncture of the Yankee Fork and the West Fork of the Salmon, half-a-dozen miles below the end of the firm's ground, where it remained idle while the search for a buyer proceeded.[48]

These were years of tight money. Finally, in 1949 Fred Baumhoff and Jack Simplot's Warren Dredging Company purchased both the dredge and the remaining ground. Crewmen brought in from outside spruced up the camp and overhauled the machinery. Operation under the new owners began in the spring of 1950 and headed upstream for known paying gravel. Just below the old mining town of Bonanza, dredgemaster Ole Johnson encountered an exposed bedrock reef, which forced the building of a dam to raise the water level high enough to float the boat over the rocky ridge to where it could dig its own pond—a tricky maneuver. Meanwhile, Simplot bought out his partner, reportedly for $75,000, and the dredge churned through a mile or so of rich ground before it reached Jordan Creek, thus exhausting the new owner's holdings. Subsequently, an agreement permitted Simplot to dredge

FIGURE 7.1. Beginning in 1940, this big modern dredge dug nearly six miles in the Yankee Fork branch of the Salmon River before becoming a museum piece. Courtesy, Idaho State Historical Society, ISHS 72-112.b.

the Morrison Claim along the creek, a job that was finished and the machine closed down in August 1952. Later, in 1953 it was discovered that the machine was still on Morrison property, so it was restarted and made a final run just far enough to get back to Simplot's ground, where it sat in its shallow pool.[49]

For the next fourteen years, the abandoned dredge sat forlornly in its final resting place about a quarter of a mile above Bonanza. Deserted, pillaged by thieves, and ravaged by vandals, it marked the end of an era. Finally, as of December 18, 1967, Jack Simplot donated the dredge and an acre of the ground on which it rests to the US Forest Service, which, with the help of dedicated volunteers, paved the way for its ultimate restoration and use as a striking museum piece harking back to the final phase of Yankee Fork's mining history.[50] A seasoned observer noted, "This artifact seemed like the epitome of the Machine in the Garden."[51]

During its lifetime, the Yankee Fork dredge produced gold valued at $1,023,024.89 and silver worth $14,298, for a total of $1,037,322.89. Sometimes skimpy records show that the cost of digging 6,330,000 cubic yards at an average of 17 cents per yard amounted to $1,076,100. For the Snake River Mining Company it was "a marginal loss," while the Warren Mining Company did slightly better because of richer ground, according to Howard A. Packard Jr. The records show that the highest profit made from the Yankee Fork dredge, Packard writes, "was [from] the wages, the increased business, and probably the taxes to the state and county."[52]

The dredge operated for eight of the dozen years between 1940 and 1952 and traversed more than six miles along the Yankee Fork, with a swathe 400 feet wide. Although a Sun Valley online guide later called the dredge "a marvel of efficiency" in its day, Arthur McGowan, who served as "gold man" on the boat, would dispute that. McGowan commented that in its lifetime the boat recovered but a fraction of the $11 million in precious metal earlier test drilling had indicated was present in the property. McGowan said that in parts of the ground, bedrock and the gold were below the machine's digging capacity. Further, he estimated that only about 60 percent of the gold that came aboard was saved. Striving for maximum yardage, bucket operators sometimes overloaded the washing plant, and much gravel would run through without getting wet. Then too, at times, especially when digging bedrock, the dredge would be tilted, which changed the slope of the fixed tables and sharply reduced gold retention. Occasionally, the buckets missed the pay streak completely, and, according to McGowan, just north of Ramey Creek, one of the richest parts of the streak was bypassed because the dredge camp had been built on top of it. Still, as late as 1962 one mining expert believed it would be possible to refit the dredge and to work the Yankee Fork profitably, even the tailings.[53]

Notes

1. Irwin Porges, *Edgar Rice Burroughs: The Man Who Created Tarzan* (Provo, UT: Brigham Young University Press, 1975), 86.

2. *Idaho Daily Statesman* (Boise), December 8, 1897.

3. Ibid., November 2, 1897.

4. *Engineering and Mining Journal* 64 (December 4, 1897): 663.

5. *Idaho Daily Statesman* (Boise), December 8, 1897; June 5, 1900.

6. Ibid., November 30, 1898; *Deseret News* (Salt Lake City), April 18, 1900; "*Boyes et al. v. Masters et al.*," *Oklahoma Reports* 17 (June 1906 and January 1907): 461–62; "*Boyes et al. v. Masters et al.*," *Pacific Reporter* 89 (April 15–June 3, 1907): 198–99; *Fourth Annual Report of the Corporation Commission of the State of Oklahoma for the Year Ending June 30, 1911* (Oklahoma City: n.p., 1911), 529.

7. *Idaho Daily Statesman* (Boise), July 24, August 14, September 12, 1899.

8. Ibid., November 22, 1900; January 1, 1901.

9. *Deseret News* (Salt Lake City), June 24, 1901; *Salt Lake Mining Review* 3 (June 24, 1901): 17.

10. *Idaho Daily Statesman* (Boise), October 1, 1898; September, 12, 24, 28, 1899; August 11, 1900.

11. Ibid., September 17, November 30, 1902; April 12 and 19, 1903; *Salt Lake Mining Review* 5 (May 15, 1903): 13.

12. *Mineral Resources, 1905* (US Geological Survey), 230; *Mineral Resources, 1906* (US Geological Survey), 230.

13. *Mineral Resources, 1907* (US Geological Survey), 296; *Idaho Daily Statesman* (Boise), June 21, 1904.

14. *Idan-ha Gold Dredge Company* (n.p., n.d.), 1, 4.

15. *Idan-ha Gold Dredge Company*, 4–6, 8, 13.

16. *Idaho Daily Statesman* (Boise), December 21, 1908; *Salt Lake Mining Review* 10 (August 15, 1908): 14; *Eleventh Annual Report of the Mineral Industry of Idaho for the Year 1909* (State Mine Inspector's Report), 52.

17. *Mineral Resources, 1908* (US Geological Survey), 422; *Mineral Resources, 1909* (US Geological Survey), 346.

18. Charles of Mining Engineers, ed., *The Mineral Industry: Its Statistics, Technology and Trade during 1911* (New York: McGraw-Hill, 1912), 792–93; *Salt Lake Mining Review* 15 (May 15, 1913): 21; *Mineral Resources, 1912* (US Geological Survey), 723; *Thirteenth Annual Report of the Mining Industry of Idaho for the Year 1911* (State Mine Inspector's Report), 49; *Fourteenth Annual Report of the Mining Industry of Idaho for the Year 1912* (State Mine Inspector's Report), 100–101; *Mining and Engineering World* 44 (January 1, 1916): 35.

19. J. B. Umpleby and D. C. Livington, "A Reconnaissance in South Central Idaho Embracing the Thunder Mountain, Big Creek, Stanley Basin, Sheep Mountain, and Seafoam Districts," Idaho Bureau of Mines and Geology Bulletin 3 (Moscow: Idaho Bureau of Mines and Geology, 1920), 16; *Engineering and Mining Journal* 113 (March 11, 1922): 423; *Mineral Resources, 1923* (US Geological Survey), 391; Stewart Campbell, *Twenty-Third Annual Report of the Mining Industry of Idaho for the Year 1921* (State

Mine Inspector's Report), 53; Stewart Campbell, *Twenty-Fourth Annual Report of the Mining Industry of Idaho for the Year 1922* (State Mine Inspector's Report), 81.

20. *Salt Lake Mining Review* 24 (May 15, 1922): 19.

21. *Spokesman-Review* (Spokane), August 5, 1928, clipping, Kimble Database, WSU.

22. *Salt Lake Mining Review* 10 (February 15, 1909): 13.

23. *Idaho Daily Statesman* (Boise), October 27, 1909.

24. *Salt Lake Herald*, August 10, 1913.

25. Quoted in *Spokane Chronicle*, October 8, 1928, clipping, Kimble Database, WSU.

26. Raoul Choate, "Geology and Ore Deposits of the Stanley Area," Pamphlet 126 (Moscow: Idaho Bureau of Mines and Geology, June 1962), 116.

27. *Engineering and Mining Journal* 142 (August 1941): 185–86.

28. *Minerals Yearbook, 1941* (US Geological Survey), 349.

29. Choate, "Geology and Ore Deposits of the Stanley Area," 105.

30. *Idaho Daily Statesman* (Boise), October 2, 1899; *Deseret News* (Salt Lake City), April 18, 1900.

31. Wayne Darlington to J. B. Hastings, Antelope, ID, October 21, 1901, copy, James D. Hague MSS, Box 11, Huntington Library, San Marino, CA.

32. Arthur Lowy to Charles Janin, New York, November 18, 1933, Charles Janin MSS, Box 18, Huntington Library, San Marino, CA.

33. Ibid.

34. *Spokesman-Review* (Spokane), May 18, 1934, clipping, Kimble Database, WSU.

35. Howard A. Packard Jr., *Gold Dredge at Yankee Fork* (Great Falls, MT: Yankee Fork Publishing, 1983), 3; H. G. Peake to Hillcone Steamship Company, "Report on the Wallace Dredging Property," December 28, 1935, copy, Yuba Manufacturing Company Papers, Meriam Library, Chico State University, Chico, CA.

36. Charles Janin to George H. Edmunds, San Francisco, May 13, 1935, copy, Charles Janin MSS, Box 8, Huntington Library, San Marino, CA.

37. H. G. Peake to Hillcone Steamship Company, "Report on the Wallace Dredging Property."

38. Packard, *Gold Dredge at Yankee Fork*, 3.

39. Kris Major, "The Yankee Fork Dredge and Its Community," *Idaho Yesterdays* 32 (Summer 1988): 23–24, 28.

40. Ibid., 24–27.

41. Ibid.; Bucyrus-Erie Company, "List of Dredges," 100075D, author's private collection; Packard, *Gold Dredge at Yankee Fork*, 7–10.

42. Packard, *Gold Dredge at Yankee Fork*, 37.

43. *New York Times*, July 28, 2008.

44. Packard, *Gold Dredge at Yankee Fork*, 37–39; Major, "Yankee Fork Dredge," 26–27.

45. Packard, *Gold Dredge at Yankee Fork*, 22, 24–25.

46. Major, "Yankee Fork Dredge," 27–28.

47. Ibid., 28–29; Packard, *Gold Dredge at Yankee Fork*, 30.

48. Packard, *Gold Dredge at Yankee Fork*, 30; Major, "Yankee Fork Dredge," 29–30.

49. Major, "Yankee Fork Dredge," 30; Packard, *Gold Dredge at Yankee Fork*, 30.

50. Packard, *Gold Dredge at Yankee Fork*, 31; Major, "Yankee Fork Dredge," 30.

51. Don Scheese, *Mountains of Memory: A Fire Lookout's Life in the River of No Return Wilderness* (Iowa City: University of Iowa Press, 2001), 64.

52. Packard, *Gold Dredge at Yankee Fork*, 34.

53. Choate, "Geology and Ore Deposits of the Stanley Area," 108; Sun Valley Guide, "A Fork of History," 2006, www.svguide.com/s06/s06_history.htm.

8

Salmon City and Leesburg

Much farther east and north of the Stanley Basin and the Yankee Fork, before the Salmon makes its big loop and swings westward, lay rich placer grounds along its shores and those of its tributaries in the Salmon City and Leesburg area. As early as January 1893, suction dredges were touted in these Lemhi County towns as the new efficient method of mass production along the Salmon River and its tributaries.[1] Prairie Basin, a high intermountain area thirty miles southwest of Leesburg, had gone through a small gold rush earlier, but difficult access generally limited mining except for a few hardy individuals. Among them at the dawn of the dredge age in 1897 were John Beck and C. W. Stayner, who had installed a dredge on Big Creek, a tributary of the Salmon, and were operating in what at least one observer thought was a novel way. Their Summit Valley Placer Company had fifteen men at work, probably with a suction rig, dredging up a stream of pebbles ranging in size from small shot to hens' eggs, which assayed about 50 percent tin and some gold. Each yard of dirt contained about a pound of such nuggets and was within ten feet to twelve feet of bedrock. Smelting made the metals easy to recover.[2]

DOI: 10.5876/9781607324751.c008

But the real prizes were in the Lemhi Valley itself. Streams heading near the crest of the Beaverhead Mountains flowed rapidly through narrow canyons, in the process gathering much detrital material, which was carried to the wide, slightly sloping valley of the Lemhi River where the heavier auriferous gravels were deposited along the way. Thus, rich placer grounds were formed along a number of streams that fed the river, especially the upper valleys of Napias Creek, Moose Creek, Bohannon Creek, and Kirtley Creek—all within the purview of Salmon City and Leesburg. Of the four areas, one of the richest, Napias, was largely worked out before the advent of dredging, but the other three provided fertile fields for the new technology.[3]

Placer mines in these areas had been worked by the usual hand methods by both white miners and Chinese going back to the 1860s and later by hydraulicking. In December 1892 rumors circulated in the area that a railroad was to be built and that a company was being organized in Butte and Anaconda, Montana, to bring in heavy dredging machinery and thoroughly work all the bars in the valley—in essence, to break all the ground along the Salmon from the mouth of the Lemhi to the narrows below the McNutt ranch, without interfering with farm and ranch interests.[4] Neither a rail line nor dredge boats were in the immediate future.

The first to feel the impact of dredging was Moose Creek, fifteen or so miles northwest of Salmon and reached by a rough, steep road. Here, the gravels were made up of even material, with only a few small boulders, and the beds were only twelve feet to sixteen feet thick. Beneath the paydirt lay alternate strata of sandstone and shale, interspersed with several seams of lignite.[5]

The great publicity attending the success of the newfangled bucket dredges in nearby Montana after 1895 was not lost on Salmon City miners, and the dredges became a reality in 1898, when Herman J. Reiling installed a boat of his own design on nearby Moose Creek, nine miles from Leesburg. Reiling had been the promoter and president of the company that built the first successful bucket dredge, the *Fielding L. Graves*, near Bannack two years earlier, although Samuel S. Harper was the instigator of the idea, designed the machinery, and supervised its construction.[6] Billed as "the largest gold dredging plant that has ever been constructed," the new Idaho boat was built in Chicago "of the New Zealand type," but with heavier parts of annealed steel. It was owned by the Pacific Dredging Company, of which Reiling was both president and general manager. The ground had been worked by David

FIGURE 8.1. The Pacific Dredging Company hauls a new steel spud past the hotel in Salmon City en route to Moose Creek. Courtesy, Lemhi County Historical Society and Museum Collection, Salmon, ID.

McNutt, who ran a hydraulic monitor from one month to three months each summer, depending on the availability of water, and who took out $75,000 over a twenty-year period. The company reportedly paid $150,000 for the property. By mid-October 1897 the boat had been floated and the machinery put in place, and Reiling expected to make a test run before the fall freeze. Work was completed on November 10 and the engine fired up, but the ground was already frozen, so the dredge worked its way into winter quarters.[7]

Just before the spring thaw, Reiling returned from Chicago "with a full crew of pilots, engineers, firemen and deck hands," who immediately busied themselves cleaning the boat of ice.[8] After startup in mid-April, expectations appeared to have been met. The machine worked well, profits were realized, and the Pacific Dredging Company seemed destined for a long and lucrative life. Unfortunately, the road would be a rocky one fraught with difficulties.

Although the machine saved a high percentage of the gold it dug, there were sometimes other losses. On a warm August day in 1899, Reiling and his wife, Bonnie, were the only passengers in a stagecoach running between Gibbonsville and Red Rock. In his official capacity, Reiling was carrying

$2,500 worth of gold, the recent take from Moose Creek, to Dillon. About three miles west of the Continental Divide, at Horseshoe Bend in Idaho, four masked men stopped the stage and at gunpoint stripped the vehicle and its two passengers of their valuables—including the gold—and the US mail, although they generously returned Reiling's watch because it was monogrammed. According to the local newspaper, these scruffy ruffians had been hanging around Bannack and had talked about holding up A. E. Graeter, who carried the gold for the Bannack Dredging Company. With a sheriff and posse on their trail, a chase ensued, but how fruitful it was is not clear.[9]

Mechanical problems also plagued the Pacific Dredging Company. In 1900 the chain broke and lost a bucket, and the dredge sank in thirty feet of water, where it remained for nearly a year before it was fished out at the cost of a good deal of money and labor. Sometime during this idle interlude Reiling resigned as general manager, apparently divested himself of his stock, and went on to other dredge enterprises in New Mexico and Colorado. The new owner was W. E. Jollette of Chicago, and the new manger was R. C. Beard, an expert with experience on dredges in Montana as well as Grangeville and Boise Basin, Idaho. The Moose Creek boat was up and running in mid-April 1901 and continued until September 2, when another disaster struck. One of its boilers exploded, seriously injuring four crewmen and killing dredge superintendent Charles Dunlop, who left behind a wife and five children in Salt Lake City. The machine was demolished. The boilers were new and had worked for only about four months. It was later rumored that the boiler had no safety valve and that the fireman had been called away at the crucial moment.[10]

Immediately, to show their faith in their property, the corporate officers proceeded to replace the boat that was destroyed and planned to open for the 1902 season as if nothing had happened.[11] How long the new rig worked or if operated at all is unanswerable, but in the autumn of 1903 the 3-W Mining Company—a syndicate made up men named Walker, Wilson, and Wood—leased the dredge and the ground of the Pacific Dredging Company for two years.[12] The fact that the Pacific Company was in dire financial difficulties was obvious before the termination of that contract. In January 1905 some of its creditors petitioned the courts to declare the corporation bankrupt, but after hearing arguments both pro and con, US Judge James H. Beatty denied the petition on the grounds that the evidence was insufficient to

FIGURE 8.2. The Pacific Dredging Company's bucket dredge on Moose Creek. Courtesy, Lemhi County Historical Society and Museum Collection, Salmon, ID.

prove insolvency.[13] But somehow, one faction of creditors led by John Mullen, to whom the company was heavily indebted for freight and wood, was able to take over during the crisis and operate the dredge themselves.[14]

The Mullen dredge, as the machine was called thereafter, was later described by the superintendent of another Salmon City dredge as an open-bucket line with about a two-and-a-foot capacity, although other sources claimed it was equipped with five-foot buckets. Its sluice was carried on a scow in the stern.

The dredge was steam-powered, with a separate engine for the grizzly and winches, and it burned wood until 1916, when local lignite became available.[15] Under John Mullen's supervision, it was operating successfully in 1907 near the junction of Moose and Daily Creeks for part of the season, but it prospected for another part.[16] Late in the year or early in 1908, Mullen moved the boat half a mile up the creek to ground sufficient to keep running for two more years. Despite the fact that it was too cold to work more than one shift that winter, he cleaned up more than $2,000 worth of gold in one six-day run, washing out $400 in values in the last four hours.[17] In August 1909 the dredge was reported to be "working steadily and making regular shipments." It "produced considerable fine bullion during the year."[18] The dredge remained active in 1910, and the following year, working with a crew of twelve men and an "old-style Bucyrus" rig, the Mullen holding was said to have produced a cumulative "barrel of money" for its owner, who was reported to have taken out between $15,000 and $20,000 in 1912.

At about the same time, the E.S.F. Edwards group, which owned 200 acres on Moose Creek adjoining Mullen's, was talking seriously about abandoning hydraulicking in favor of a dredge.[19] The boat worked for another year but apparently did not operate in 1914 and 1915. Mullen had a machine digging again in 1916, and it continued until at least the 1920 season.[20] At that late date, John Mullen of Salmon was listed in the *American Mining Manual* as president and general manager of the Moose Creek Company, operating a five-and-a-half-foot steam-powered dredge with fifteen workmen.[21] That information and the semi-official nature of the publication would tend to add credence to the idea that the dredge had been upgraded to one with a larger capacity, perhaps prior to the 1916 season. Records of neither the Bucyrus Company nor the Yuba Manufacturing Company show a delivery of a new bucket machine to Mullen or his firm,[22] but that would not preclude the possibility that they bought used equipment. In any event, John Mullen had a long and profitable run.

The Kirtley Creek placers would also eventually become strong producers. Located a few miles southeast of Salmon, considerable gold had been taken out by hydraulicking during the early 1890s before operations stopped. In 1902 it was announced that 1,000 acres along the stream of the "best dredging ground in Idaho" were under option to a Chicago group, which expected to work that ground on a larger scale. Experts hired by the new investors estimated that the property might yield $8,000 per acre.[23]

For whatever reason, the deal fell through, as did another dredge installation at nearby Leesburg proposed by the Napias Placer Mining Company in 1909.[24] A year later a San Francisco dredging company bonded ground owned by the Lemhi Orchard and Irrigation Company, invested $20,000 into carefully prospecting it, then bought "some hundred" acres for $50,000 or $75,000—the press could not agree on the price.[25] On this property the deposits were relatively new gravels containing shot-sized gold in the upper part of the basin and flake gold in the lower basin. Incorporated under the name Kirtley Creek Gold Dredging Company, the new firm was headed by John Martin of Ross, California, who was both president and chief owner. The officers quickly ordered a nine-foot hybrid Bucyrus-Yuba bucket dredge at a cost of $148,022 to be delivered on-site. Apparently, Bucyrus built the machine and Yuba modified it to fit the specifications of the Hammon Engineering Company, which had been chosen to supervise the Idaho operation where Nelson Dickerman was in charge.[26] A University of California graduate, Dickerman had been with the Yuba Consolidated Gold Fields dredge for one year and had served as assistant to general manager Norman Cleavland of Natomas Consolidated for two years, both companies operating in California.[27]

Since the boat was to be powered by electricity furnished by the Lemhi Light and Power Company, local workers began cutting and distributing poles for the six miles of wiring in March 1911, a few weeks before the first consignment of machinery was due to arrive. L. Hohl, a veteran dredge manager and now consulting engineer for the new Kirtley Creek firm, appeared on the scene at about the same time, along with two experts from the Yuba Manufacturing Company, to oversee the building process. The shipment of machinery was expected to weigh about 1,000 tons and would be carried in thirty railroad cars over the newly completed link with the Oregon Short Line at Armstead, Montana.[28] By early May, most of the electrical supplies were on the ground and ten cars of lumber had arrived—massive pieces of Oregon fir measuring sixteen-by-twenty inches and eighteen-by-eighteen inches, some as long as seventy feet. A month later Doc Hudkins, who had a contract to haul material by wagon from a railroad station to the dredge site, had completed transfer of the machinery but was still moving the big timbers. Fifty people were at work on various aspects of preparation, which included building a two-story hotel for crewmen and cottages for the superintendent and foreman. The plant also included a complete machine shop, with air compressors and a crane able to lift twenty tons. The pit had been dug, and actual construction would soon begin, with a completion deadline of August 1. Installation proceeded as "rapidly as unlimited capital and skilled workmen can push it," according to the Boise *Statesman*, with owners of nearby placer ground on Bohannon Creek and Geertson Creek displaying a keen interest.[29]

While completion of the dredge took a month longer than the schedule called for, when Hohl left for his home in San Francisco in early August, he promised that the boat would be afloat that same week and would be operational by September 1. As was usual for most dredges of its time, the machine had a wooden hull. Its sixty buckets would handle 150,000 cubic yards of dirt per month. Electrically powered, its various motors required a total of 500 horsepower to run the plant. It came with a double bank of gold-saving tables, one of which was soon removed. The dredge was also equipped with a system for heating the rim to enable it to dig frozen gravel in winter weather—an innovation that proved unworkable. With a crew of twenty men divided into three shifts, the Kirtley Creek boat was soon running successfully.[30] After a number of profitable weekly cleanups, toward the end of November it completed its first monthly cleanup "and the results were exceedingly flattering," in the words of one observer.[31]

In the long run, the Kirtley dredge's experience was fairly typical of the industry in its adolescent days. By this time, many of the kinks had been worked out of the equipment, and experienced engineers knew how to drill properly and to accurately predict future values of the ground. The Salmon River and its tributaries were different from the Snake; their bars were deeper and better defined and their gold coarser and more plentiful. The boat was under Dickerman's supervision until August 1913, when he left to take charge of a dredge on the Nechi River in Colombia, owned by the Oroville Dredging Company, a British firm. His successor on Kirtley Creek was B. H. Liskow, a University of Michigan graduate in mechanical engineering with experience in California dredging.[32]

Both men were competent managers who knew their business and were responsible for maintaining their company's strong reputation as a consistent producer. *Mineral Resources of the United States* for 1915 gave a well-earned tribute: "Aside from the work of the Boston-Idaho Co., in Boise Basin, this is the most important dredge operation in the State." In June of that year, when twenty-six automobiles loaded with Idaho Falls Elk Club members made their annual trek to their salmon bake in Lemhi County, a visit to the Kirtley Creek dredge was included.[33] The boat ran most of the time from its maiden voyage in 1911 until early 1918, when it was dismantled and shipped to Marysville, California. The company's "ample and well equipped machine shop," together will all of its tools, was bought by the Drilling Development Company, another local concern.[34]

The trials and tribulations of the Kirtley Creek machine were typical; hard wear and tear often led to downtime—the dredger's expensive bugaboo. The stern gantry broke and caused twenty-eight days of delay in 1914, the ladder winch went out and cost twelve idle days in 1915, and a screen tread failed, which meant five more lost days that same year. Electric power had many advantages but was not always dependable. For one reason or another, power outages totaled twenty-three days in 1912, nineteen days in 1913, and thirty-six days in 1914. Ice on the digging ladder caused several wrecks of the ladder until steam pipes to prevent icing were added. The dredge dug at an average depth of sixteen-and-a-half feet on slightly sloping terrain, making it difficult to keep floatation because of shallow ground and grades, not to mention a good deal of clay to be handled at bedrock.[35] But despite delays and setbacks, the Kirtley Creek boat had a good long run and produced 24,145 ounces of gold and 2,038 ounces of silver.[36]

Located about a dozen miles east of Salmon City, Bohannon bar, on the creek of the same name, was another area of Lemhi County with great potential for dredging. Chinese had worked the creek bed in the 1870s, but it was not until 1895 that limited mining with hydraulic giants began, using water stored in a large reservoir and carried some distance by flume. In 1912 only one nozzle was operated on the bar, with an annual recovery of about $17,000. Here, the auriferous gravels occurred in three benches with thicknesses up to fourteen feet south of the creek and up to thirty-one feet north of it.[37] As early as 1911, while the Kirtley Creek boat was still under construction, eastern owners of placer ground watched and considered following suit. Others soon viewed the Kirtley machine as "the forerunner of at least two equally large dredges" to be installed in the near future, one on Bohannon Creek and one on Geertson Creek not far away.[38]

Enter Colonel Gustave Pabst of Milwaukee, president of the Pabst Brewing Company. Like other sharp businessmen, Pabst was constantly diversifying his assets and already had investments in ranchland near Salmon City, as well as mines in Colorado and Arizona. In 1910 various claims on Bohannon bar were consolidated and acquired by the Pabst Brewing Company. After prospecting the ground thoroughly with a Keystone drill, Pabst organized the Bohannon Dredging Company, ordered a six-foot Yuba dredge, and appointed F. H. Krueger of Milwaukee superintendent in charge.[39] This was not a new machine. Earlier, it had dug at Hammonton, California, for the Yuba Dredging Company until deeper digging boats were built; it had then been sold to Pabst, who put it in the hands of the Yuba Manufacturing Company for dismantling, shop overhauling, and shipping to his ranch, where it was reassembled by Yuba mechanics in 1913. Unofficially, it was said that the renovation cost $84,000.[40] All of this was done without the considerable newspaper fanfare that had

FIGURE 8.3. Colonel Pabst's renovated six-foot Yuba during reconstruction near Salmon City. Courtesy, Lemhi County Historical Society and Museum, Conley Collection, Salmon, ID.

accompanied the coming of the Kirtley Creek machine. Like its Kirtley counterpart, the Bohannon Creek dredge was electric and purchased current from the Lemhi Light and Power Company for seventy-seven cents per kilowatt hour. With a wooden hull and close-connect bucket line, it was originally built to dig to fifty-five feet, but it was remodeled to reach only forty feet.[41]

Superintendent Krueger bowed out almost as soon as the Bohannon dredge started up in 1913 and was replaced by Fred De Land, who had been in charge of its on-site construction.[42] That first winter disclosed a serious weakness when the bow gantry broke down under the strain of heavy digging. The gantry had to be redesigned and reconstructed, with rebuilding of much of the hull to provide bracing in the stern. The cold weather was too severe. At times, frozen ground made digging impossible and caused caving, which endangered the dredge; at other times, the eight-mile-long ditch that carried water vital to floatation and pumping froze. On occasion, the grizzly and tables would freeze solid.[43]

FIGURE 8.4. Pabst's dredge with rebuilding complete. Courtesy, Lemhi County Historical Society and Museum, Conley Collection, Salmon, ID.

There was another set of problems of a less technical nature. One was the old struggle over control of water. At least once, the company was engaged in a lawsuit over that question in a case involving Bohannon Creek, which it lost, but details are lacking.[44] More engaging was the fight over title to the land. The company claimed 2,080 acres of placer ground. The bars on both sides of the stream were level and the gravels covered with enough topsoil to make them desirable for cultivation using creek water for irrigation. Prior to installation of the dredge, mining had been confined to the lower bars or those next to the creek. Higher bars could not be mined at all until the dredge arrived. For many years this ground was covered by the placer claims of the company or its predecessors without patenting them, merely by doing annual labor on them. Without patents, these lands appeared on the records of the US Land Office as still vacant and open to settlement. Thus, a good many homestead and desert land entries had been made on lands covered by the Bohannon bar placer claims. This led to litigation between the mining company and various claimants for the land under agricultural entries. Tradition held that prospecting and discovery proved the mineral character of land and as such it could not be open for farming.[45]

It fell to the US Land Office to handle the cases. In so doing, it broke each 160-acre placer claim into 10-acre units and passed on each one. The results of such hearings were often favorable to the Bohannon Dredging Company, but at times they were also frustrating and expensive. Some 10-acre tracts awarded to the company as mineral were surrounded or joined by units deemed non-mineral; thus, when moving the boat to one of its legitimate 10 acres, the firm might have to cross a tract set aside for farming to which it had no right. The company bought several of these agricultural claims and in 1917 was endeavoring to acquire more so it could move its machine freely across the bar. During these negotiations it temporarily suspended its dredging operations and turned to hydraulic mining with two Hendy seven-inch monitors. The owners of a good deal of land on Geertson Creek, not far from Bohannon bar, faced the same troubles. The Lemhi Orchard and Irrigation Company still held considerable ground but had changed its original mission. Deciding to go into mining on its own, it test-drilled part of its property, found it promising, and prepared to install a dredge, only to find itself in litigation with agricultural claimants. Even a new Lemhi power plant was planned to provide the needed electricity.[46] The boat on Geertson Creek was never forthcoming.

Despite such distractions, the Bohannon Dredging Company did manage to operate over a period of several years. According to various issues of *Mineral Resources*, it began in 1913, probably in September; dug for seven months during 1914; dug for six months in 1915, when it handled more than 500,000 cubic yards of material; and was again listed as active in 1916.[47] That was probably its final fling. Its name did not subsequently appear in any of the publications.

In the fall of 1930 the dredge still sat silent in the field where it last closed down. At that time, it was owned by Walter McLaren of Dayton, Ohio, who was anxious to sell. McLaren commissioned George Holmes, a veteran California expert, to examine the machine in September. Holmes found that the buckets were still good but that the lower tumbler needed replacement and that spare parts were lacking and perhaps impossible to obtain.[48] No sales followed, but two years later Holmes passed on the gist of his report to Charles Janin after McLaren learned that Janin was investigating placer ground for an English company and had inquired about the old Bohannon boat.[49]

During the Depression of the 1930s, the raising of the price of gold rejuvenated the gold-dredging industry. In late 1937 two Reconstruction Finance

Corporation mining engineers, Arthur V. Corry and Charles Bielenberk, both of Butte, were in Salmon with William J. Porter of Spokane to examine Moose Creek property of the Placer Exploration Company near Leesburg with respect to a federal loan. Porter was president of the firm, which since that spring had been test-drilling and shafting the ground preparatory to installing a large dredge. The company's holdings in the area were large and it was "highly pleased" with the test results,[50] but there is no indication that the firm's loan application was ever accepted.

In 1941 the Idaho-Warren Dredging Company's four-foot Marion bucket-line boat on Hughes Creek in the Gibbonsville district operated during the last quarter of the year. In 1942, in a truncated season in which the War Production Board closed down gold mining, the dredge recovered 1,900 fine ounces of gold and 123 ounces of silver. After the ban was lifted on July 1, 1945, it was nearly two years before the rig was manned and started up again; it worked from April 20 to December 30, 1947, and again until February 5, 1948. At that point it was dismantled by the Warren Dredging Company and moved to the Elk City district in Idaho County.[51]

The first mention of the use of a dragline dredge came in 1939 in the Mackinaw district, but no owner was recognized. Smith Brothers was working on the Hagel ground two years later with a dragline and was the leading gold producer in the district. In 1942 the Smiths were back at the Hagel ranch, where they recovered 2,466 fine ounces of gold and 203 ounces of silver. After the interlude for World War II, the brothers returned in 1946 with a new partner, incorporated as Smith Brothers and Courtis, Inc. The year 1947 was a good one, still at the Hagel ranch, but the dragline was idle through the following year. In 1958 another dragline owned by Washington Iron Works was also operating on Hagel-owned ground, but it was in the Kirtley Creek district, not Gibbonsville.[52] An era of gold dredging was over in the area.

Notes

1. *Mining and Scientific Press* 66 (January 14, 1893): 21.

2. *Idaho Daily Statesman* (Boise), June 4, 1897.

3. Joseph B. Umpleby, "Geology and Ore Deposits of Lemhi County, Idaho," US Geological Survey Bulletin 528 (Washington, DC: Government Printing Office, 1913), 51–52.

4. *Idaho Daily Statesman* (Boise), December 17, 1892.

5. *Engineering and Mining Journal* 103 (January 13, 1917): 95–96.

6. Ibid. 64 (December 4, 1897): 663; *Mines and Minerals* 19 (March 1899): 380.

7. *Mines and Minerals* 19 (March 1899): 380; *Mining and Scientific Press* 77 (October 15, December 3, 1898): 382, 558; *Salt Lake Tribune*, October 10, November 28, 1898.

8. *Salt Lake Tribune*, April 5, 1899.

9. *Alder Gulch Times*, August 25, 1899; *New York Times*, August 24, 1899.

10. *Idaho Daily Statesman* (Boise), May 13, 1901; *Dillon (MT) Examiner*, September 4, 1901; *Deseret Evening News* (Salt Lake City), September 5, 1901; unidentified note, Charles Janin MSS, Box 43, Huntington Library, San Marino, CA [hereafter Charles Janin MSS]; *The Locomotive* (Hartford, CT) 22 (September 1901): 181; *Salt Lake Mining Review* 2 (March 15, 1901): 12.

11. *Dillon (MT) Examiner*, September 4, 1901; *Idaho Daily Statesman* (Boise), January 1, 1902.

12. They were L. D. Walker, Robert Wilson, and J. A. Wood. *Idaho Daily Statesman* (Boise), November 17, 1903.

13. Ibid., January, 5, 24, 25, 1901.

14. B. H. Liskow to Charles Janin, Salmon, ID, September 13, 1915, Charles Janin MSS; *Mineral Resources, 1911* (US Geological Survey), 595.

15. *Mineral Resources, 1911*, 595.

16. *Idaho Daily Statesman* (Boise), November 17, 1907; *Mineral Resources, 1907* (US Geological Survey), 350.

17. *Salt Lake Tribune*, January 31, 1908.

18. *Idaho Daily Statesman* (Boise), August 30, 1909; *Mineral Resources, 1909*, 350; *Mineral Resources, 1911*, 595.

19. *Idaho Daily Statesman* (Boise), August 7, 1911; *Mineral Resources, 1914* (US Geological Survey), 605; Umpleby, "Geology and Ore Deposits of Lemhi County, Idaho," 52.

20. This information comes from the annual *Mineral Resources of the United States* for these years.

21. *American Mining Manual 1920* (Chicago: Mining Manual Co., 1920), 186.

22. I have depended on the Bucyrus Company's "List of Dredges," 100075A–100075F, author's private collection, and Samuel Johnston, *"Gold Dredges": The Part Played by the Yuba Manufacturing Company in the Great Romance for Placer Gold, Platinum and Tin* (San Francisco: John Henry Nash, 1927), both of which list all new dredges and to whom they were sold during this period.

23. *Salt Lake Telegram*, November 1, 1902.

24. *Salt Lake Mining Review* 10 (January 30, 1909): 12.

25. *Engineering and Mining Journal* 90 (December 24, 1910): 1276; *Salt Lake Herald*, December 11, 1910; *Idaho Daily Statesman* (Boise), December 12, 1910.

26. *Engineering and Mining Journal* 90 (December 24, 1910): 1276; Yuba Manufacturing Company Estimate for 9 ft. Electric Driven Placer Dredge, September 26, 1910, copy, Yuba Manufacturing Company Papers, Meriam Library, Chico, CA. Both Yuba and Bucyrus list the Kirtley Creek dredge on their production lists. Bucyrus Company's "List of Dredges," 100075C; "List of Dredges Built by Yuba Manufacturing Company," typescript, no pages, copy, Charles Romanowitz Papers, Rasmuson Library, University of Alaska, Fairbanks; *Mining and Engineering World* (Chicago) 44 (January 1, 1916): 35; *Idaho Daily Statesman* (Boise), August 7, 1911.

27. *Bulletin of the American Institute of Mining Engineers* 96 (December 1914): xvii.

28. *Idaho Daily Statesman* (Boise), March 25, 1911; *Salt Lake Mining Review* 8 (April 15, 1911): 15.

29. *Idaho Daily Statesman* (Boise), May 22, June 24, August 7 and 19, 1911.

30. Ibid., August 7, 1911; Charles Janin, information for the Kirtley Creek Gold Dredging Company (ca. 1915), Charles Janin MSS.

31. *Idaho Daily Statesman* (Boise), November 25, December 31, 1911; *Salt Lake Mining Review* 13 (November 15, 1911).

32. *Idaho Daily Statesman* (Boise), September 11, 1911; Frederick M. Halsey, *Investments in Latin America and the British West Indies,* Bureau of Foreign and Domestic Commerce, Special Agents Series 169 (Washington, DC: Government Printing Office, 1918), 301, 519.

33. *Mineral Resources, 1915* (US Geological Survey), 547; *Idaho Daily Statesman* (Boise), June 28, 1915.

34. *Salt Lake Mining Review* 20 (January 15, 1919): 21; Bucyrus Company's "List of Dredges," 100075C; *Idaho Daily Statesman* (Boise), January 13, 1919.

35. Janin, information for the Kirtley Creek Gold Dredging Company; B. H. Liskow to Janin, Salmon, August 2, 1915, both in Charles Janin MSS.

36. G. J. Orris and J. D. Bliss, "Geologic and Grade-Volume Data on 330 Gold Placer Deposits," US Geological Survey Open-File Report 85–213 (Menlo Park, CA: US Geological Survey, 1985), 1.

37. Umpleby, "Geological and Ore Deposits of Lemhi County, Idaho," 52–53, 122.

38. *Salt Lake Mining Review* 8 (September 15, 1911): 8; *Salt Lake Telegram*, February 26, 1912; *Mineral Resources, 1912* (US Geological Survey), 728.

39. *Engineering and Mining Journal* 103 (January 13, 1917): 95–96.

40. This is listed by Yuba Manufacturing under dredges that "have [been] redesigned and rebuilt." Johnston, *"Gold Dredges,"* 42; George L. Holmes to Charles Janin, San Francisco, September 26, 1930, Charles Janin MSS.

41. *Engineering and Mining Journal* 103 (January 13, 1917): 95; Charles Janin, information on Bohannon bar placers, n.d., Charles Janin MSS.

42. *Idaho Daily Statesman* (Boise), September 19, 1913.

43. George L. Holmes to Charles Janin, San Francisco, September 26, 1930; Janin, information on Bohannon bar placers, n.d., both in Charles Janin MSS.

44. *"Bohannan Dredging Company, a Corporation, et al., Appellants, v. James G. England et al., Respondents," Idaho Reports* 30 (October 1917): 721–27. This case deals with the appeal of the lower court's decision not to give the dredging company a new trial after it had received an adverse verdict in an "Action to determine priority of water rights."

45. *Engineering and Mining Journal* 103 (January 13, 1917): 95–96.

46. Ibid., 96–97; *Idaho Daily Statesman* (Boise), May 30, 1913; July 21, 1914.

47. *Mineral Resources, 1914*, 605; *Mineral Resources, 1915*, 547; *Mineral Resources, 1916* (US Geological Survey), 589.

48. George L. Holmes to Charles Janin, San Francisco, September 26, 1930, Charles Janin MSS.

49. George L. Holmes to Charles Janin, San Francisco, October 20, 1932, Charles Janin MSS.

50. *Spokesman-Review* (Spokane), October 17, 1937, clipping, Kimble Database, WSU.

51. *Minerals Yearbook, 1941*, 352; *Minerals Yearbook, 1942*, 381; *Minerals Yearbook, 1947*, 1409; *Minerals Yearbook, 1948*, 1503.

52. *Minerals Yearbook, 1939*, 331; *Minerals Yearbook, 1941*, 352; *Minerals Yearbook, 1942*, 381; *Minerals Yearbook, 1946*, 1427; *Minerals Yearbook, 1947*, 1409–10; *Minerals Yearbook, 1948*, 1503.

9

South and Middle Forks of the Boise River

In the autumn of 1899, the Montana Gold Company located several hundred acres of ground on the South Fork of the Boise River in Camas County, much farther east than Junction bar. H. L. Childs was representing this group; during the summer he tested the ground and had it staked off preparatory to putting a dredge on it. In 1902 there was a Childs Dredging Company "with several good claims" at some unidentified place on the South Fork, but it was probably in the same locale.[1] In neither case was there any activity.

At about the same time, Charles T. Brant was spokesman for the Scottish American Development Company, which was trying to promote investment in a placer property at Junction bar (modern Featherville) in Elmore County during the late 1890s. Junction bar was located at the convergence of the Feather River and the South Fork of the Boise. Brant's firm had 320 acres of ground, which he described as "about the richest he ever heard of in the state." The gravel averaged 28 feet in depth and 23 cents to the yard in gold values. A crew of six had been working during the fall of 1898, prospecting with a drill said to be "a perfect piece of machinery [whose] handling

DOI: 10.5876/9781607324751.c009

cost $1,500." Brant gave the impression that the Scottish company initially purchased 120 acres of its ground for $10,000 and acquired the rest by location and relocation. The relocated property had once belonged to Frank Crossman, who not only made open threats against the firm, but he and a friend named McClure were arrested twice for abortive attempts to blow up the drill with explosives. Both times they were discharged by a sympathetic justice of the peace. Brant noted that more than $400,000 had been taken out of Junction bar by the old-timers and their primitive methods. That amount only touched the surface. His company planned to use the most modern technology—a large dredge.[2]

In mid-December Brant returned from Denver, headquarters of the Scottish company, where he met with the directors. He reported that the board had approved funds not only to complete purchase of the ground but also to buy the dredge to be installed in the spring of 1899.[3] The machinery would arrive as soon as the roads were open, and it was hoped that construction could begin in about a month, Brant said in late March 1899.[4] By mid-May, with no dredge in sight, William Orr of Denver appeared on the scene representing the London Gold Extraction Company to consult with some of the owners of placer ground on which the Scottish American Development Company had taken options. Brant no longer represented the latter firm. Orr indicated that the dredge was on its way to Junction bar. Months later, although no longer their agent, Brant declared that he had negotiated a deal for 120 acres under bond to the Scottish company, that a prospecting outfit had been ordered from Chicago, and that if the ground proved as valuable as anticipated, "a dredge will be put in next spring."[5] Brant did not mention the backers of the transaction, but it must be inferred that his plans fell through.

In 1905 several other abortive attempts were made to install dredges at Junction bar. In September 1904 E. E. Burlingame announced that as a result of his investigations of ground there, he was convinced that Junction bar was destined to be "the center of the most extensive dredging operations in the country." A former treasurer of the New England Guaranty and Investment Company in Providence, Rhode Island, Burlingame now assumed the same position in the new Elmore Gold Extraction Company, formed to exploit the area's rich placers. President of the firm was H. C. Harle of New York, treasurer of the Empire Light and Power Company. Other officers and board members included three local residents of Elmore County's county seat,

Mountain Home: L. B. Green, a leading local lawyer; W. A. Reynolds, cashier of First National Bank of Mountain Home; and W. J. Turner, both mayor and postmaster of the town.

With nominal capital of $500,000 in shares of $1 each, Elmore Gold Extraction set out quickly to enrich all of its shareholders. A friendly piece in the *United States Investor* laid out the formula. First, the company controlled sixteen claims of 20 acres each on the South Boise, in an environment said to have produced $100 million in gold within a radius of 15 miles of its property, $9 million of which was taken out just across the river. In addition, the ground was ideal for dredging, with a gravel bed about 30 feet deep, and the ground averaged values of $3 a cubic yard. The company was having a modern dredge built, capable of handling 3,000 yards a day, and the makers of this machine "guarantee to save over 90 per cent of the free gold values at a working expense of not over five cents a yard." Using the figures and based on an operating season of 250 days, the firm's spokesman foresaw an annual income of $2.25 million, but to be safe he also calculated that should the ground turn out to contain only 50 cents a yard, the annual return would still be $375,000, enough to pay dividends of 75 cents on every share of capital stock. It was hoped that the dredge would be up and running by August 1905.[6] That did not happen. Figures of the value were much too high to be realistic. The Elmore Extraction Company had all of the earmarks of one of the scams conceived by sharp operators to empty dredge shareholders' pockets.

That same year the Cleveland Gold Dredging Company, with offices in the Citizens Bank Building in Cleveland, Ohio, issued a thirty-six-page illustrated prospectus extolling the virtues of "rich gold placer ground, situated on the South Boise River, at Junction Bar, Elmore, County, Idaho." At about the same time, R. J. Fitzgerald of San Francisco, who was involved in dredging in California, organized a syndicate to operate in the same area. The Fitzgerald people had options on 1,500 acres on both sides of the river and brought in a large drill similar to the one used on Junction bar ground the previous year. Then, too, the Boston and Idaho Gold Dredging Company, already active in Boise Basin, also had claims at Junction bar and was at work on them.[7] There is no evidence that any of the groups was able to follow through at this time. No real dredge would appear at Junction bar for a number of years.

Since 1909, at least 320 acres of ground in the same district had been owned by Captain Rupert Winters, who in 1898 had installed and operated one of the early dredges in the Boise Basin for the Boston and Idaho Gold Dredging Company. Under the name Elmore Placer Company, which he controlled, Winters had attempted to work the placer on the South Boise with a combination hydraulic and dredge system. With bedrock forty feet down, a large pit was dug and a stationary steam-driven three-foot bucket line installed to raise gravel from a large pit to wash through long sluices. Although Winters established the value of the ground, his water supply was so limited that he could only run his machine about three months of the year, so he was forced to abandon it in 1914. Winters spent the next seven years trying to interest enough capital to fund a major dredging enterprise. In 1919 it seemed he had succeeded when it was announced that the Keystone Dredging Company of San Francisco had acquired 800 acres along the South Boise and Feather Rivers and that Winters was involved. Keystone proposed putting three dredges on the ground, but something went awry at that point and the firm did nothing.[8]

Finally, in 1921 the South Park Dredging Company took up Winters's challenge, incorporated his land into its larger holdings, and entered the field in earnest. This was a powerful outfit, with a good deal of money and highly skilled management behind it. Heading up the firm was Wendell Hammon, the first successful pioneer in California gold dredging, a key figure in the organization of the Yuba Manufacturing Company—a major producer of bucket-dredge equipment—and an investor in a number of dredging operations. In a proposal signed by both parties in December 1921, Yuba Manufacturing agreed to act as the South Park Dredging Company's agent and supervise the purchase and construction of a nine-foot electric dredge capable of digging to thirty feet. Yuba would provide the complete machine, at an estimated net price of $276,900, including a profit of 12 percent of the net cost, not to run more than $30,770. South Park would furnish the pit, provide water for flotation, provide the electricity, and pay all taxes on the machinery or parts thereof. The completion date was set at June 1, 1922. When the final score was settled, the invoice South Park received gave the certified cost of the dredge—*Yuba No. 51*—as $280,656.98. With the maximum profit to the manufacturer added, the total figure came to $311,426.98.[9]

For a time at least, several public sources, including the *Engineering and Mining Journal,* insisted that the South Park Dredging Company was to build two dredges, which did not prove to be true. By the end of 1921 Yuba had shipped fifty freight cars of machinery and supplies to Mountain Home, the closest rail point to Featherville. As was not unusual in the high country, severe winter weather inhibited both the freighters who moved the massive consignment the remaining forty-five miles and the fifty to sixty workers employed to assemble the big dredge. At the same time, the company was also constructing an electric transmission line from the Idaho Power Company facilities in Mountain Home to the Featherville dredge site. South Park either owned or had options on two miles on the Feather River and one mile along the South Fork of the Boise—a total of 1,150 acres of ground. It was reported that the firm would invest $1 million in the enterprise, half of that before it dug a single bucket of gravel.[10]

The new Yuba started digging near Featherville on August 21, 1922, "an important date for Elmore County and the State," according to Idaho inspector of mines Stewart Campbell, who proceeded to give a brief history of the decline of the rich placers and famous lode mines in the area. The surface mines were worked out early; most of the ground suitable for hydraulicking went next; and the quartz mines were closed because of litigation or a lack of power, with production dwindling "to practically nothing." A general malaise had settled over mining. The coming of the South Park Dredging Company heralded what Campbell hoped was the beginning of recovery for the industry.[11]

The *Idaho Daily Statesman* heartily echoed these sentiments and shortly after startup gave a detailed description of the big machine, apparently forgetting for the moment that Winthrop Estabrook had brought a much larger one, a 15-footer, into the Boise Basin in 1910. In vivid prose, the *Statesman* depicted every phase of the operation, beginning with a picturesque paragraph: "At the junction of the Feather and Boise rivers, in Elmore county, a gold-hungry monster is greedily devouring hundreds of tons of gravel a day. Sixty-seven huge buckets, the claws of the monster, feed the material which they have scooped from the river bed, into the Monster's teeth at a rate of 20 buckets per minute. And each bucket contains, on the average, nearly seven cents worth of gold!" According to the reporter, the 9-foot dredge could move about 150,000 cubic yards per month, and it was estimated that it would take six years to mine the ground for which it had been built. The ground averaged about 20 cents a cubic yard. At that rate, the monthly return could be figured at about $30,000, or approximately $2.25 million for the boat's six-year life. Including stops for greasing and repairs, in a normal day's work the machine could handle 5,000 cubic yards. If that gravel were loaded into horse-drawn wagons with a 1-cubic-yard capacity each, allowing 30 feet for each wagon, they would form a procession 28 miles long: projected over the anticipated life span of six years, the procession would reach around the world two times.[12]

For its time, *Yuba No. 51* was a state-of-the-art gold dredge. The 105-foot hull was built entirely of Oregon fir. Its buckets weighed nearly a ton apiece; the digging ladder was 75 feet long and weighed roughly 50 tons; the tailing stacker was 100 feet in length; and the entire dredge weighed close to 200 tons. Everything was electrically driven: 520 horsepower was distributed to the various components from the winch room.[13]

Beginning in the late summer of 1922, the South Park boat worked steadily until June 1928, except during 1926, when it dug for only six months. Its output of gold was valued at more than $200,000 in 1923, when it was the second-largest producer in Idaho—a ranking it repeated in 1925 and again in 1928. In 1924, 1926, and 1927 it led all mines in the state in gold recovery.[14] Overall figures are not available, but it is clear that it was a profitable enterprise.

Another area targeted by dredgers was Little Smoky Creek, which eventually flowed into the South Fork of the Boise River. In her day, Ada M. Hollister was a very unusual woman. Not only was she was an investor in mining property, she was actively engaged in the process of planning and development. In May 1908 it was reported that Mrs. Hollister of Stanford, Connecticut, had gone to Hailey to look after her various mining interests in Idaho.[15] In early January 1911 she was featured in the Boise *Statesman*. Now a resident of the state, living in a cabin on Little Smoky Creek in Blaine County, this "charming woman, still in the prime of life," was the owner of a gold-producing property and a coal mine. She had discovered the mine herself. Seven years earlier she had purchased the old Newton Revis placer ground, 640 acres in all, on Little Smoky and until six months previous had operated both businesses personally. Revis had worked the placer in a primitive fashion but still took out a good deal of gold. Intent on doing better, Ada Hollister spent considerable time testing the ground to determine the best mode of working it. She brought in Rupert Winters, former superintendent of the Boston and Idaho Gold Dredging Company, as a consultant. Winters's report strongly recommended that a bucket

FIGURE 9.1. Typical piles of tailings, produced near Featherville by the dredge of the South Park Dredging Company, in the background. Courtesy, Idaho State Historical Society, Boise, ISHS 72-112.b.

dredge be installed on Little Smoky Creek, which flowed across the length of the property.[16]

With that in mind, Hollister organized the Little Smoky Mining Company, with Arthur T. Kingsbury of Akron, Ohio, as president and herself as vice president. Kingsbury was currently district agent of the Equitable Life Assurance Society. A manager was to be appointed as soon as the dredge was ready. A crucial problem was that the money to buy the machine had yet to be raised through the sale of stock, which was why Hollister was calling on the publisher of the Boise newspaper in January 1911.[17] Nothing further developed from this proposal, so it can be inferred that no adequate capitalization materialized.

A year later M. M. Tolle, former superintendent of the Eclipse Mining Company in Hailey, asked the Yuba Manufacturing Company for estimates on a new three-and-a-half-yard dredge, to be located sixteen miles from Fairfield, Idaho. This sounds very much like Little Smoky. Yuba responded with a quote of $71,412 for a standard-type steam-operated machine, with a wooden hull, revolving screen, and stacker ladder.[18] There was no follow-through on Tolle's part and no indication that he was linked to Hollister's abortive effort.

Not until nearly another quarter of a century had passed was serious consideration again given to dredging on Little Smoky Creek. This time the dream became reality, and a business firm was created. The site was no longer in Blaine County; a new Camas County had been created by the legislature in 1917. Whether it was the same piece of ground Hollister had hoped to dredge is not clear. Further, we know little about who owned the machine.

This new outfit, the Little Smoky Dredging Company, started up its bucket dredge on Little Smoky Creek, near Fairfield, in 1935 and immediately became the largest gold producer in the county during that and the following year, even though it operated only during part of 1936 and its output was less than half that of 1935. In 1937 the boat was idle throughout the year.[19] Subsequently, a year later the Baumhoff-Fisher Company installed a four-foot bucket-line dredge on Little Smoky Creek but discontinued it in 1939 and acquired the two-and-a-half-foot dredge of the Little Smoky Dredging Company. The company moved the dredge to its Warren ground, where it commenced digging on August 25, 1940.[20]

The placers of the Middle Fork of the Boise proved even more difficult to handle. In general, they were more difficult to prospect and often capricious where sustained production was concerned. Some early finds did prove profitable, especially those of Boise King Placers between Swanholm and Pfeiffer Creeks below Atlanta, but even these placers would not sustain complicated ditch-work and hydraulicking. Not until mid-June 1941, when Boise King Placers brought in a seven-foot boat, was there an effort to dredge the river.[21]

The newcomer was Yuba Manufacturing Company's dredge, *Yuba No. 153*. As it usually did, Yuba sent an engineer, in this case Edward A. Foster, to the site for a thorough inspection of the machine and "to put it through its paces." Apart from some adjustment of the lower spud keeper and the fact that the heating system was not yet complete, Foster pronounced it a "workmanlike dredge," not likely to cause much trouble. The dredge had run for three hours before on two different occasions, with a crew of three experienced men training a group of green hands.[22]

The boat continued to operate in 1942 near Twin Springs, recovering 6,340 fine ounces of gold and 2,149 ounces of silver in a run that ceased on October 15, in accordance with the War Production Board's ban. The output of the Middle Fork district fell to a single ounce for 1943. Boise King Placers never returned when the war was over. Instead, the Pyramid Mining Company took over its property and its bucket dredge in 1946 and dug the ground from April 30 to September 30 but did not come back the following season. In all, the seven-foot boat is estimated to have taken out $200,000 in gold before it was done.[23]

Notes

1. *Idaho Daily Statesman* (Boise), October 16, 1899; January 1, 1902.
2. Ibid., September 20, November 15, 1898.
3. Ibid., December 11, 1898.
4. Ibid., March 30, 1899.
5. Ibid., May 18 and 20, August 25, 1899.
6. *United States Investor* (Boston, New York, and Philadelphia) 16 (March 18, 1905): 400.
7. *Idaho Daily Statesman* (Boise), October 29, 1905; advertisement in Bartleby's Books, Washington, DC, http://bartlebysbooks.com/shop/bartleby/56992; WorthPoint, 2007, auction, http://www.worthpoint.com/worthopedia/1905-06-gold-mining-dredging-junction-bar.
8. Unidentified clipping, 1921?, Scrapbook, Yuba Manufacturing Company Papers, Meriam Library, Chico, CA; *Idaho Daily Statesman* (Boise), August 27, 1922; Robert N. Bell, *Twelfth Annual Report of the Mining Industry of Idaho for the Year 1916* (State Mine Inspector's Report), 11; *Electrical World* (New York) 73 (February 15, 1919): 351.
9. Yuba Manufacturing Company Proposal to South Park Dredging Company, September 1, 1921, signed by South Park officials, December 2, 1921, copy; E. Adlahide, Auditor, Certification of Costs, September 1, 1921, copy; Invoice to South Park Dredging Company, September 30, 1922, copy, all in Yuba Manufacturing Company Papers, Meriam Library, Chico, CA.
10. *Engineering and Mining Journal* 112 (December 31, 1921): 1074; *Engineering and Mining Journal* 114 (July 1, 1922): 44; unidentified clipping, 1921?, Scrapbook, Yuba Manufacturing Company Papers, Meriam Library, Chico, CA; *Idaho Daily Statesman* (Boise), March 5, 1922; August 7, 1972; Stewart Campbell, *Twenty-Third Annual Report of the Mining Industry of Idaho for the Year 1921* (State Mine Inspector's Report), 58; Stewart Campbell, *Twenty-Fourth Annual Report of the Mining Industry of Idaho for the Year 1922*, 86.
11. *Twenty-Fourth Annual Report of the Mining Industry of Idaho for the Year 1922*, 84, 86.
12. *Idaho Daily Statesman* (Boise), August 27, 1922.

13. Ibid.; *Twenty-Fourth Annual Report of the Mining Industry of Idaho for the Year 1922*, 86.

14. *Mineral Resources, 1928* (US Bureau of Mines), 668; *Mineral Resources, 1929* (US Bureau of Mines), 390–91.

15. *Mining Science* (Denver) 57 (May 7, 1908): 448.

16. *Idaho Daily Statesman* (Boise), January 12, 1911.

17. Ibid.; *Akron Official City Directory Supplemented by Directories of Barberton and Cuyahogo Falls* (Akron, OH: Burch Directory, 1911), 671.

18. *Mining Reporter* 52 (October 5, 1905): 343; Estimate of the Yuba Manufacturing Company, December 10, 1912, copy, Yuba Manufacturing Company Papers, Meriam Library, Chico, CA.

19. *Minerals Yearbook, 1935* (US Geological Survey), 207; *Minerals Yearbook, 1936* (US Geological Survey), 395; *Minerals Yearbook, 1937* (US Geological Survey), 321.

20. *Minerals Yearbook, 1938* (US Geological Survey), 351; *Minerals Yearbook, 1939* (US Geological Survey), 325; *Minerals Yearbooks Review of 1940* (US Geological Survey), 345.

21. *Minerals Yearbook, 1941* (US Geological Survey), 349.

22. Edward A. Foster to W. B. Macaulay, June 6, 1941, copy; Foster to Macaulay, San Francisco, June 16, 1941, copy, both in Yuba Manufacturing Company Papers, Meriam Library, Chico, CA.

23. *Minerals Yearbook, 1942* (US Geological Survey), 365, 371; *Minerals Yearbook, 1943* (US Geological Survey), 371; *Minerals Yearbook, 1946* (US Geological Survey), 1440–41; *Minerals Yearbook, 1947* (US Geological Survey), 1408; Merle W. Wells, *Gold Camps and Silver Cities*, Bulletin 22, 2nd ed. (Moscow: Idaho Department of Lands, Bureau of Mines and Geology, 1983), 64.

10

The Yukon Dredge

Located in the Idaho Panhandle, seventy miles east of Spokane, the Coeur d'Alene region has long been recognized as the state's premier mining area, much better known ultimately for its sustained production of silver, lead, and zinc than for placer gold. Still, it was the discovery of the precious metal that touched it all off. The first frenzied rush into the region was in response to the cry of "gold" in 1883, after A. J. Prichard discovered deposits of the glittering metal on Eagle and Prichard Creeks—streams that drained into the North Fork of the Coeur d'Alene River. Thousands flocked into the new town of Murray, but the "city a half a mile long" strung along Prichard Creek was soon eclipsed by the finding of a rich variety of other ores near the South Fork of the river, the development of which in time would lead to the exploitation of one of the world's greatest mining districts. A report issued in 1989 gave the round figures of the area's production since lode mining commenced ninety-five years earlier: more than a billion ounces of silver, 8.5 million tons of lead, 3 million tons of zinc, and large quantities of antimony, cadmium, copper, and gold.[1]

DOI: 10.5876/9781607324751.c010

Placer mining continued on the creeks with traditional labor-intensive techniques and some hydraulicking, although the first dredge appeared at a surprisingly early date. In 1886 J. F. Cargill of Minneapolis represented a firm that owned exclusive rights in Idaho and Montana for the sale or operation of a patented gold dredge for deep placers. That spring, when he visited the Coeur d'Alene district, he succeeded in signing up one local company to introduce one of the machines and expected several more to follow. He did not name the firm or where it operated. Cargill was ahead of his time when he used the phrase "[these dredges] will work a revolution in placer mining."[2] Eventually that would be true, but not yet.

When the mining engineer-editor Thomas A. Rickard visited Prichard Creek more than thirty years later, he noted "three monuments to unsuccessful dredging" nearby, close to the mouth of Trail Gulch. Two of them were derelict Risdons manufactured in San Francisco. Both had started on easy ground, he said, but both owners "had neglected to ascertain whether there was any gold in the gravel." One of the dredge-masters subsequently turned to agriculture and "today reaps a richer harvest from the gold of the cornfield," according to Rickard. The third dredge dug along the gulch for half a mile, not always reaching bedrock, but it soon sank: "The hull with its shingled roof could be seen from the road, resting in a pond that gave back the blue of perfect sky."[3] Rickard gives no dates or other details, but an intriguing notice in 1898 by a news correspondent writing about Eagle City, once a thriving center of an earlier placer boom but now defunct, promised a new technology that would restore that town to at least a measure of its former glory. The answer lay in the "scheme now in process of development to dredge the deep placers" of the area." All of the machinery was on the ground, and a contract for the necessary lumber had been let to the Raven sawmill for delivery as road conditions permitted.[4] It is possible that, when completed, this may have been one of Rickard's ill-fated "three monuments," although it may have been too early for it to have been a Risdon bucket dredge this far from California.

In 1903 one gold dredge was operating in Shoshone County, that of the Delta Gold Mining Company, and the following year one unidentified boat was listed in *Mineral Resources*.[5] According to *Mining American*, three dredging companies in the county paid taxes that year: the Delta Gold Mining Company, the Northern Gold Company, and the Mascot Gold Mining

Company. None paid the net profit levy, but each paid an "improvement" tax, which ranged from $418 to $426. Which one of the companies was working was not noted, but it is clear that none of them was making money. In 1905 the Mascot Gold Mining Company's dredge was listed as active. Little information is available about the firm, except that the company was incorporated in 1902, with nominal capital of $500,000 and headquarters in Wallace. Directors were H. E. Chaney of Missoula, Montana; C. S. Crysler, J. M. Savage, and J. F. Griffin, the latter three from Delta, a town close to Murray. Delta was near the spot where Trail Creek joined Beaver Creek, which in turn flowed into Prichard Creek—all streams with considerable placer mining history.[6] Before the end of 1883 it became a center of some importance and remained so for a few years, as long as the placers along the creeks and their gulches remained profitable. In the 1890s the mines played out and Delta became a virtual ghost town. The crowning blow came in 1902 when the town site was acquired by the Mascot Gold Mining Company, which put a dredge on the site. The buildings were torn down, and eventually all that remained were heaps of tailings.[7]

Federal documents in 1905 commented that since 1898, placers in the area "were practically exhausted, the hydraulic method [of] operation being superseded by dredges," and that since 1904 one of the two dredges digging near Murray had moved to near Pierce in Nez Perce County. Also in 1905, the Boise *Statesman* reported that engineer Albert Burch was one of the principal owners of a "gold ship" that had been moved from the Coeur d'Alene country the previous fall and installed near Sumpter in Oregon, where the dredge operated and turned a profit before it closed down for the winter. Apparently, success was short-lived and the machinery of the Burch-Burbidge boat, as it was called, was shipped back to Idaho, to Warren, where it was converted from steam to electricity.[8] That was probably the last dredge machine in the Coeur d'Alenes for a few years. More than a decade would pass before *Mineral Resources* would list another in the area.

When big-time dredging finally came to Prichard Creek, it was the Guggenheim Exploration Company that was responsible. Beginning in 1906 and working through a subsidiary, the Yukon Gold Company, the powerful Guggenheims would build a small dredging empire, with seven boats in the Klondike and several more in Alaska. Finally, in 1916 the Yukon Gold Company took over three new gold-dredging properties in the Lower

Forty-Eight—two in California and one in Idaho. These new placer grounds were acquired through leasing agreements, with an estimated gross gold content of $5.35 million, based on further development, and net profits of roughly $2 million to the company without further work. In California, on the North Fork of the American River, the company gained river claims and leases with an anticipated return of about $1 million. The second California acquisition was a group consisting of the Trinity Exploration and Carr properties. In Idaho the property was that of the Coeur d'Alene Mining Company, on Prichard Creek near Murray, with a total area of about 8.5 million cubic yards and an estimated gold content of about $1.35 million.[9]

The Coeur D'Alene Mining Company was a Boston concern, with nominal capital of $1.25 million and lawyer Barton E. Eames as president and John F. Murphy as managing agent. The firm had long owned 2,800 acres of placer ground, extending seven miles west and four miles east of Murray as well as on Eagle Creek, plus a number of side gulches along Prichard Creek. This property had been idle for ten or fifteen years until 1917, when it was leased to Yukon Gold. The agreement between the two companies provided that the Coeur D'Alene Mining Company would receive 25 percent of net profits after expenses and commissions had been paid and 40 percent of net profits after all costs of installation, plant, and everything else had been met.[10]

The No. 9 dredge of the Yukon Gold Company, working on Elderado Creek outside Dawson, was another hybrid; its machinery was built by Bucyrus of Milwaukee, with special finishing touches and assembly by Yuba Manufacturing Company. One of the first of the all-steel dredges, it was hauled 3,000 miles to its site in the Klondike, where it began digging in 1911. After five years its ground was exhausted, and the dredge received a new assignment in Idaho. It was then dismantled and shipped down the Yukon River by barge 2,000 miles to St. Michael, where it was reloaded onto a steamer and carried another 2,500 miles to Seattle. The final leg of its journey was by rail flat cars to within seven miles of Murray, where the well-traveled machine arrived sometime in mid-1917.[11]

On Prichard Creek, the boat was reassembled about three-quarters of a mile below the town by a number of experts, including Yukon Gold's general manager Oscar B. Perry, who dropped by while on leave from his command of the 27th Engineers Regiment in France, made up entirely of mining specialists. The dredge was built on a slope and launched sometime in November into the pond previously created as an old hydraulic elevator pit, a reminder that all "modern" technology might not work on the creek. The final tweaking of machinery allowed it to begin operations on December 6, under the supervision of A. B. Hughes—a veteran Canadian dredge man—with a force of fifteen, split into two nine-hour shifts. Before it were fourteen miles of ground along the creek that had been prospected the previous summer.[12]

The seven-and-a-half-foot Yukon boat was electrically powered, with its 6,600-volt current from Shoshone County Power stepped down to 440 volts for use onboard. In studying the ground, company engineers discovered that the gold was concentrated mainly close to bedrock thirty feet or so down. Bedrock consisted of slate, with ribs of hard quartzite, and was generally considered too difficult to dredge, as the three relics mentioned by Rickard seemed to testify. Thus, the Yukon people modified their No. 9 boat especially for the Prichard Creek ground. It was strengthened wherever possible, and the use of manganese-steel was expanded beyond such parts as tumblers, buckets, and replaceable lips that received excessive wear to components on which it had not been previously used. Another basic change was to add larger electric motors. Whereas a 100 horsepower motor was sufficient to drive the bucket line in the Klondike, in Idaho 150 horsepower was required. The result was the ability to dig two feet into bedrock if the panner determined that the gold had penetrated that far.[13]

"Plenty of steel" and a more powerful dredge was part of the secret to success, as were the able, innovative, and experienced dredge men who set the pace. Hughes soon gave way to a new engineer in charge, D. H. Ferry, a Stanford University graduate who was also a seasoned alumnus of the Canadian school of dredging. Ferry would be in charge during the entire eight-and-a-half years of the boat's working life, with headquarters in what had been the courthouse when Murray was the county seat and with the former jail used as a storehouse for company supplies. Roy Darr had been a winchman for seven years on dredge No. 2 in Dawson, and the dredgemaster, Isaac Hinkle, had also honed his skills in the North. All of the crewmen, especially Ferry and Hinkle, were charged with keeping the wheels turning and the boat in continuous operation. Downtime was money lost. During the first nine months of operation on Prichard Creek, the Yukon dredge handled over a million cubic yards, with a running time averaging 86 percent. During the first eight months of 1919, its yardage had increased by 25.8 percent, with a 14 percent decrease in cost in the face of rising prices. This was accomplished by sharp management, attention to details, calculated avoidance of delays, and keeping a ready supply of spare parts on hand. The boat was equipped with a small machine shop, which included an air compressor and oxy-acetylene welding and cutting equipment. The latter had recovered its cost thirty times in the first five month of operation, according to Ferry, largely because of its ability to cut manganese-steel, something no ordinary tool could do.[14]

One change to equipment was dictated at least indirectly by the nature of the gold recovered. Unlike the Snake River, Prichard Creek's showed no fines, and the black sands contained no value. Gold on the creek was coarse or sometimes in thin flat pieces. Thomas Rickard was shown a nugget weighing nearly five ounces. Cleanups were made every ten or fifteen days. In between those intervals, after the dredge first started up, more than one employee was tempted to pocket a few choice coarse pieces from the gold-saving tables. After one miscreant was caught in the act and sent to jail, manager Ferry saw fit to enclose the tables with woven wire, with only the panner, or so-called gold man, entrusted with a key to the locked door.[15] Parenthetically, the Yukon dredge, like most others of its kind, collected not only precious metal but also small mementos of another sort: fragments of silver spoons, vintage coins from as far back as the 1860s and 1870s, and, while digging across the claim—said to be the stronghold of a local desperado—nearly a thousand spent bullets of all kinds.[16]

There is no doubt that Ferry ran a tight, efficient ship. That is not to say that the dredge had no problems with its component parts. The spud—the long steel post that was dropped through a socket to hold the dredge in position against the digging thrust—had failed five times in fifteen months before a new type was devised of plate-girder construction, twenty-four inches by thirty-six inches and forty-eight feet in length, of which eighteen feet was in the ground. The new version was twice as strong as the old but still under a terrific strain. Stepping forward was done by raising the spud and winching cables firmly anchored onshore. The dredge men learned that by reducing the length of each forward advance by one-third, that yardage was increased; although the boat had to be moved more frequently, digging from top to bottom with a smaller bite was done more quickly. Moreover, the shorter step facilitated better cleaning of the bedrock and lessened the severe shocks caused when a bucket struck large rocks.[17]

Although the dredge lost time because the spuds broke in 1918, it operated through the winter except for minor stoppages but was forced to close down for two weeks for additional repairs in June the following year. Still, from the beginning it was regarded as a highly successful operation that was producing between $80,000 and $90,000 a month. It was even rumored that the Yukon Gold Company might dismantle two more of its boats when they ran out of ground and send them to Prichard Creek—something that did not happen.[18] But at about the same time, the company sent engineers to the Elk City district of Idaho to take options on a large block of land along both the Elk and American Rivers and ordered drilling equipment to be moved in. In the US Senate hearings to discuss strategic minerals during World War I, it was argued that the stream was rich in tungsten and that the Yukon dredge could be diverted to recover scheelite, if needed.[19]

In the summer of 1919 an issue of the *Spokane Chronicle* showed a picture of the boat and carried the headline "Will Soon Rip through City Streets." A separate article in the same issue was more explicit: "Yukon Dredge Will Tear up City Streets." The company's leased ground followed the creek through Murray and for some distance beyond. Oscar Perry, chief engineer of all of Yukon's dredges, was visiting at the time and expressed pleasure with the operations on Prichard Creek; the boat had already scraped more than a mile of bedrock, with handsome returns. The firm had sufficient ground remaining to keep the machine running for a number of years, and when that stream was finished, he expected similar work to be carried out on Eagle Creek and other tributary gulches.[20] Oddly, there seems to have been no negative reaction to the fact that the huge machine was about to enter the town site and dig the pay streak under the streets. By the end of 1919, Idaho inspector of mines Robert Bell was extolling the company as "one of the best financed and managed dredging enterprises the State has had."[21]

Before plans could be completed, disaster struck in the spring of 1920. The dredge had closed down for annual repairs and overhauling. It had been leaking some while digging, which was not uncommon, but the leaks had ceased when the dredge was idle. The two watchmen on duty the night before were working on the hopper early in the morning when they heard things falling on the deck. The boat listed sharply, and they barely made it to shore before it went down in twenty-five feet of water. There was no time to use the emergency pump, and an electric device in the stern designed to give an

FIGURE 10.1. The Yukon Gold dredge devours part of the town of Murray on Prichard Creek, 1922. Courtesy, Idaho State Historical Society, Boise, ISHS 72-112.b.

automatic warning of water did not function because water came into the bow. No cause could be assigned until a diver inspected it and possibly until the dredge was refloated. Company spokesman estimated it would take two weeks to raise the vessel and three more to make repairs.[22]

Several divers were brought in from the Pacific Coast, but six weeks passed before the dredge could be raised, at which point D. H. Ferry predicted it would be at least two weeks before it would dig again. Cofferdams had to be built so men could work below the surface. Once again, the massive steel spud had to be rebuilt. At the same time, the company also installed a new tailings conveyor belt 185 feet long and made repairs to the screens, pumps, and stacker. The boat did not resume operations until April 3, 1921.[23]

When it did resume operations, it spelled doom for the waning town of Murray. Early in 1922, having reversed its field and returned to the town site, the big steel monster chewed up a substantial part of the town. All of the Prichard Creek frontage and all of Gold Street were dredged, with many of the buildings torn down and replaced by ugly windrows of rocky tailings.[24]

While the company did not divulge its profits, both Idaho state mining inspector Robert N. Bell and the US Geological Survey's compilers of the annual *Mineral Resources* regularly mined the reports of both the company and the county assessor for yearly operating statistics. Bell reported that the gold output of Shoshone County for 1918 was $302,000 and that the major portion probably came from the Yukon dredge. Bell's guess was probably a good deal higher than it should have been. During 1920, according to the county assessor, the dredge dug 1,295,000 cubic yards and produced a total of $96,332, no doubt below its normal average because of so much downtime, but its net profit was still $16,683. A preliminary report by the US Geological Survey for 1921 indicated that while the value of all metals mined in Idaho had dropped more than 50 percent since 1920, gold output was up by $11,500. That year the boat on Prichard Creek recovered values of $149,821, an average of 9.726 cents per cubic foot, at a cost, excluding depreciation, of $90,668. Unofficial figures for 1922 showed that the company grossed $124,028 but that after expenses, profits were only $20,083. The bumper year was 1923, when Yukon Gold produced $244,803 in gold, with an operating gain of $156,740. By the end of that year, the boat had worked the creek for three miles—two miles west and one mile east of Murray—and was still continuing its run.[25] In 1924, when it and the South Park Dredging Company in Elmore County were Idaho's main producers of placer gold, Yukon's operating profit was $67,712, but that figure was down to $29,071 in 1925, its last full working season. The big electric machine dug during the first five months of 1926 and recovered $40,000 before closing down permanently.[26]

The US Bureau of Mines estimated that over its active working life since 1917, the Yukon dredge had produced approximately $1.14 million in gold. Howard W. Staples, who had served as assistant manager on the machine for four years, quotes a figure of $1.27 million taken from Prichard Creek before its last cleanup on May 29, 1926. How much of that amount was profit? According to the *Salt Lake Mining Review* three years later, profits amounted to exactly $500,431.20 since startup.[27] Where the editor of the august *Review* found those figures is not recorded. They sound high, but in the absence of other numbers, they will have to stand.

Dismantling of the machinery commenced immediately, with the various motors going to the Coeur d'Alene Hardware and Foundry Company of Wallace and most of the rest of the equipment to a company on the Rogue

River in Oregon. The hull was left in the last pool where it had dug, a relic of a past mining era, to be viewed by visitors as late as the 1930s.[28]

In the 1930s a dragline operated briefly in the area, along Trail Creek. Later, in the 1940s, at least two new concerns showed an interest in Prichard Creek. In the spring of 1940 the Porter Brothers, Helena, Montana, dredge operators, leased and bonded 2,800 acres of patented land below the ground cleaned by the Yukon boat earlier. A few years later the Silver Dollar Mining Company leased placer ground on an unidentified portion of the stream.[29] There is no indication that either group followed through with its plans. For all practical purposes, the Yukon Gold Company was the only firm successful in dredging Prichard Creek.

Its Idaho operation was just one part of that major corporation's widespread dredging activity. At one time, general manager Oscar Perry had charge of properties not only in Shoshone County but also in California, Nevada, British Columbia, Yukon Territory, and Alaska. Both hydraulic equipment and dredging equipment were used. After World War I, Perry returned to manage the company's very successful tin operations in Malay, where four large American-type dredges were installed.[30] No other multiple dredge corporation of that size and scope ever ran a bucket dredge in Idaho.

Notes

1. Robert Higgs, "Coasian Contracts in the Coeur d'Alene Mining, District," June 2, 2004, Independent Institute, www.independent.org/publications/working_papers.

2. *Helena (MT) Daily Herald,* April 17, 1886.

3. Thomas A. Rickard, *The Bunker Hill Enterprise* (San Francisco: Mining and Scientific Press, 1921), 26.

4. *Idaho Daily Statesman* (Boise), March 12, 1898.

5. *Mineral Resources* (US Geological Survey), 1904, 187; *Mineral Resources* (US Geological Survey), 1905, 240; *Mineral Resources, 1914* (US Geological Survey), 604.

6. *Mining American* 50 (December 22, 1904): 677; *Mineral Resources, 1905,* 240; *Idaho Daily Statesman* (Boise), September 30, 1902; *Mining World* 29 (December 5, 1908): 843–44.

7. Robert Wayne Smith, *History of Placer and Quartz Gold Mining in the Coeur d'Alene District* (Fairfield, ID: Ye Galleon, 1932), 69–70.

8. *Mineral Resources, 1905*, 238; *Idaho Daily Statesman* (Boise), October 19, 1905; Brooks Hawley, *Gold Dredging in Sumpter Valley* (Baker, OR: Baker Printing and Lithography, 1977), 17.

9. *Copper Curb and Mining Outlook* (New York) 14 (April 4, 1917): 37.

10. Sidney Norman, ed., *Northwest Mines Handbook* (Spokane: Northwest Mining Association, 1918), 1:54; *Moodys Manual of Railroads and Corporation Securities (Twenty-Third Annual Number, Industrial Section)* (New York: Poor's, 1922), 2:1318.

11. Bucyrus Company, List of Dredges, 100075C, author's collection; Rickard, *Bunker Hill Enterprise*, 28.

12. *Alaska and Northwest Mining Journal* (Seattle) 11 (September 1917): 8; *Alaska and Northwest Mining Journal* 11 (November 1917): 137; *Engineering and Mining Journal* 104 (December 8, 1917): 1145; Rickard, *Bunker Hill Enterprise*, 29.

13. Rickard, *Bunker Hill Enterprise*, 29.

14. Robert N. Bell, *Twentieth Annual Report of the Mining Industry of Idaho for the Year 1918* (State Mine Inspector's Report), 85; Stanford University, *Alumni Directory and Ten Year Book 1891–1920*, 3 (Palo Alto: Stanford University, 1921), 598; Rickard, *Bunker Hill Enterprise*, 28–30.

15. Rickard, *Bunker Hill Enterprise*, 32–33.

16. Ibid., 32.

17. Ibid., 31; *Engineering and Mining Journal* 106 (October 5, 1918): 645.

18. *Salt Lake Mining Review* 20 (February 28, 1919): 36; *Engineering and Mining Journal* 107 (June 21, 1919): 1103; *Spokane Chronicle*, June 17, 1919; April 15, 1921, clippings, both in Kimble Database, WSU; *Alaska and Northwest Mining Journal* 12 (February 1918): 32.

19. *Spokesman-Review* (Spokane), October 3, 1919, clipping, Kimble Database, WSU; *Minerals and Metals for War Purposes Hearings before the Committee on Mines and Mining, United States Senate, Sixty-Fifth Congress, Second Session, on the Bill H.R. 11259* (Washington, DC: Government Printing Office, 1918), 159–60.

20. *Spokane Chronicle*, August 22, 1919, clipping, Kimble Database, WSU.

21. Robert N. Bell, *Twenty-First Annual Report of the Mining Industry of Idaho for the Year 1919*, 116.

22. *Engineering and Mining Journal* 109 (June 19, 1920): 1377.

23. *Seattle Chronicle*, July 1 and 7, August 12, 1920; April 15, 1921, all clippings, Kimble Database, WSU.

24. Smith, *History of Placer and Quartz Gold Mining in the Coeur d'Alene District*, 61–62.

25. Rickard, *Bunker Hill Enterprise*, 33; *Idaho Daily Statesman* (Boise), January 10, 1922; *Moodys Manual of Railroads and Corporation Securities (Twenty-Second Annual No., Industrial Section)* (New York: Poor's, 1921), 2:842; *Mineral Resources, 1920* (US

Geological Survey), 255; *Mineral Resources, 1921* (US Geological Survey), 415; *Mineral Resources, 1922* (US Geological Survey), 239; *Mineral Resources, 1923* (US Geological Survey), 97; *Spokesman-Review* (Spokane), December 10, 1923; *Engineering and Mining Journal* 117 (May 24, 1924): 869.

26. *Mineral Resources, 1924* (US Bureau of Mines), 277; *Mineral Resources, 1925* (US Bureau of Mines), 540–41; *Mineral Resources, 1926* (US Bureau of Mines), 444.

27. *Mineral Resources, 1926*, 444; Smith, *History of Placer and Quartz Gold Mining in the Coeur d'Alene District*, 102–3; *Salt Lake Mining Review* (January 15, 1929): 39.

28. Smith, *History of Placer and Quartz Gold Mining in the Coeur d'Alene District*, 102–3.

29. Wakefield Dort Jr., "Gold-Bearing Gravels Near Murray, Idaho," Idaho Bureau of Mines and Geology Pamphlet 116 (Moscow: Idaho Bureau of Mines and Geology, August 1958), 17; *Engineering and Mining Journal* 141 (April 1940): 78.

30. *Engineering and Mining Journal* 146 (November 1945): 121.

11

Warren

Twenty-seven miles southeast of Florence, gold was discovered on the South Fork of the Salmon River in the spring of 1862 by a party headed by James Warren, who has been described as a "shiftless individual, a petty gambler, miner and prospector."[1] Warren's Diggings, as the new camp was originally called, quickly blossomed into a boisterous town, although the presence of the Californians who replaced some of the riff-raff of early Florence made it more orderly than many new mining centers.[2] At an altitude of about 5,900 feet above sea level, Warren was much like the other high placer regions of Idaho County: sub-zero temperatures and heavy snow isolated the camp in the winter, and even during the rest of the year, travel was arduous and expensive. In 1900 Waldemar Lindgren wrote that it "is one of the least accessible mining camps in the West, being about 130 miles by wagon road from the nearest railroad."[3]

Located in the Salmon River Mountains, the Warren district was set in a broad basin surrounded by lofty summits roughly 8,000 feet in elevation, except for the drainage outlet cut by Meadow Creek. The lowest part of the

DOI: 10.5876/9781607324751.c011

basin near its center was Warren Meadows, an extensive alluvial plain, which also ran broad belts up various tributaries of the main stream. It was the shallow gold-bearing gravels of Meadow Creek that touched off the rush. As in the case of Florence, much of the placer ground was covered with several feet of soil or black peat, but the stream made rich and easy pickings for early prospectors.[4] The camp grew steadily until its numbers reached 1,500 in 1865, then it dropped off to a third of that number two years later. For the time being, the cream had been skimmed, as evidenced by the fact that Chinese miners were admitted in 1872. According to Lindgren, who visited Warren in 1897, much of the production in the previous twenty-five years had been done mainly by the hard-working Chinese.[5]

The first public rumblings of possible mechanization in the area appeared in 1894, when the six joint owners of 480 acres of virgin placer ground on Secesh Creek, about a dozen miles from town, announced their hopes to develop their property.[6] They had never gone down to bedrock, their spokesman said, but they had prospected enough to know that it was rich. The location was flat, and success would require a dredge. There was no follow-up, so apparently they were unable to raise the funding they needed to proceed.

Six months later one of the owners, E. B. True, bought the William B. Knott mine from former governor Norman B. Willey. Willey was a mining entrepreneur, who came to office when Governor George Shoup resigned in late 1890. Willey had called out the Idaho National Guard and also secured federal troops to quell the bloody labor dispute in the Coeur d'Alenes two years later.[7] After a brief two-year term, the former governor had resumed his work buying and selling mining property. Willey, George Riebold, and several others owned 200 acres of Warren placer ground, which was sold to Philadelphia capitalists for $26,000 in late January 1895. Some of the tract had been worked and found to be very rich. Riebold pointed out that the Chinese who mined the placers just down the creek had taken out $2.5 million over the previous ten years. The new owners had purchased a "dredging machine," which was then at Pocatello and would be shipped when spring weather allowed. Riebold also noted that Warren was snowbound, that groceries were scarce. He and others had snowshoed fifty-two miles to Salmon Meadows (later New Meadows) en route to Boise.[8] Reports from Warren near the end of July indicated that machinery for the dredge was now arriving—that two more carloads were at Weiser, awaiting shipment to the site.[9]

The "dredging machine" turned out to be a Marion steam shovel. Precisely when it was installed and started up is not clear, probably sometime in late 1895. Two years later the Warren Placer Mining Company was operating the shovel with a crew of forty-four men on Meadow Creek a mile-and-a quarter below town, near the mouth of Steamboat Creek. This was not dredging in a conventional sense; rather, it was a pit operation of a Rube Goldberg nature. From the pit, the shovel loaded the gravel into cars, which were then hoisted by steam over an inclined plane and dumped into long sluice boxes, supported by a trestle thirty feet above the level of the valley. About 600 tons of dirt could be handled per day, but only the lowest ten feet were considered rich enough to wash. The gold value of the washed gravel was said to run at forty cents a cubic yard. In 1897, the plant ran from May 1 to November 1.[10] In all likelihood, it was not active the next year. Like its early counterpart in Florence, the plant was deemed a failure because of the high operating costs and low yardage handled.[11]

Not until the summer of 1905 did another dredge come to Warren. At that time, a small secondhand Risdon was brought in from Weaver Basin, California, by a man named Alexander, probably with financial support from Risdon Iron Works. The machine dug for two months in the fall of 1905 and proved to be in poor shape and much too light for the work at hand. During the following year it sank and was abandoned.[12]

In the summer of 1907 two California engineer-promoters, Charles Munro and J. S. Wyatt, also had their eyes on the Warren Meadows ground Alexander owned. In trying to entice Jonathan Bourne Jr., a mining investor and future senator from Oregon, to help fund further testing and development of the property, the pair was highly laudatory: "It is the most attractive property we have seen outside of the big California dredging fields, and is the first property [we have] ever visited that exceeded our expectations." Bourne said "no, thank you," and immediately pulled the rug from under them, with instructions to discharge all employees, dispose of supplies, cut every possible expense, and dispose of the drill.[13]

Munro and Wyatt then brought in a third partner, J. S. Wallace, and in the spring of 1908 they signed a three-part agreement. Wyatt would devote his time to securing options and conducting other necessary business connected to the Warren Meadows dredging ground. Munro and Wallace would finance the expenses of living and traveling while putting together the company's promotional proposition, with any profits or bonuses on options to be divided as follows: Wyatt and Munro 40 percent each and Wallace 20 percent. In the preliminary report, drawn up by Munro, emphasis was on the need for a large seven-foot bucket dredge that would cost $150,000 installed, a power plant for another $50,000, the placer ground at $179,000, and prospecting at an additional $10,000. The report predicted that these expenditures would provide a return of $389,700 in the first two years and pay 29.8 percent on the investment for the next fourteen years. These were honorable men. Munro was a highly respected graduate of the University of California, with a dozen years' experience managing dredges, and his figures were realistic.[14] Times were hard in the aftermath of the financial panic of 1907, however, and there were no takers.

One of the unused by-products of the Warren placers that was left behind in the tailings was monazite sand. Technically, that mineral could be broken down to produce thorium nitrate, which was essential to the manufacture of incandescent gas mantles in the United States but which had long been imported from Germany. When World War I erupted, Germany's export of magnesite was cut off, and American manufacturers of such lighting equipment cried out in dismay. At that point, in early 1915, negotiations were under way for a thorough sampling and investigation of the big gravel deposits of Warren Meadows, with the intent of introducing large dredges designed to capture both gold and magnesite.[15] No action was forthcoming. Indeed, more than a decade and a half elapsed before dredging of any sort became a reality in the district.

From time to time, the state inspector of mines chided mining investors for not taking advantage of Warren's rich ground. In 1911, for example, during a time of stagnancy, Robert N. Bell noted that there were still 400 or 500 acres in the lower Warren Creek Basin that were ideal for a big dredging enterprise. "This beautiful stretch of meadow land," he said, was covered with a bed of gravel about sixteen feet deep, had no sizable boulders, and carried average values of ten cents to fifty cents per cubic yard. In addition, this superb tract was "probably one of the richest resources of monazite sand that exists anywhere in the United States," in such a form as to be readily separated by an easy method.[16]

Again in 1919, when the area was still in the doldrums, Bell's *Annual Report* carried the same message, arguing that although the district was believed

FIGURE 11.1. This boiler was en route to one of the early dredges in Warren. Courtesy, Idaho State Historical Society, Boise, ISHS 72-112.b.

to have already produced a total of $15 million from its placers, there were still several hundred acres of prime dredging land two miles below the town of Warren. Nor did Bell fail to mention the presence of monazite as an "attractive by-product of the dredging process."[17] To no avail: lean years for Warren prevailed.

Although small placer mining and quartz mining continued throughout the period, the total value of all gold produced in the Warren district in the twenty-seven years from 1902 to 1928 inclusive was $21,581.30, an average of slightly over $830 per year. A mere $300 worth of gold from placers was recovered in 1929; $315 was the take in 1930. Then, in 1931 placer production began to rise dramatically, increasing in the next five years to a total of $1,593,062.[18] The dredges had arrived. Coupled with other factors like the rising price of gold—ultimately, in 1935—improved new small dredges, and more modern dragline equipment, the gold boats were largely responsible for pulling the mining industry out of the doldrums in the Warren district.

The first sign of activity along those lines came in a press release that circulated widely in midsummer 1931. It was an announcement that "the California Dredging Company" was building a big dredge on Warren Meadows, just below town, "causing Warren to boom again, not like the old gold rush days, but in a surer and more safe way."[19] The name was wrong—it was the Idaho Gold Dredging Company—but it was a Californian who was the moving figure in the operation. The son of a German immigrant, Fred Baumhoff developed a trucking business over Echo Pass in the Sierras, then in the early Depression years he migrated to Idaho, where he would have a highly successful career in gold dredging.[20]

The Idaho Gold Dredging Company was his initial venture into Idaho mining. He partnered with Elza Thomas Fisher, more familiarly known as "Al" or "Kid." Little is known about Fisher's previous mining experience. It would be interesting to know what role Al Fisher may have played in the E. T. Fisher Mining Company, a Seattle firm that operated successfully under that name for eight consecutive seasons in Wyoming. From 1933 to 1940 inclusive, the firm ran a dragline on Rock Creek in the Atlantic City district.[21]

However that may have been, the exact source of financing of the Baumhoff-Fisher-Warren partnership is not clear, but observers mention "associates," and the company started small. In 1931 the firm dismantled a dredge in Grant County, Oregon, hauled it by truck, and rebuilt it on Warren Meadows, where it commenced digging on September 11 the same year. It was a wood-hulled machine powered by steam. Two Lentz boilers produced about ninety horsepower each and required about two-and-a-half cords of wood per day.[22]

Initially, the company was operating on a group of placer claims turned over to it by Andrew Kavanaugh, who also owned other ground in the area. This ground was within what would prove to be the Warren district's most productive region—a belt about five miles wide that ran seventeen miles or so south from the Salmon River. In 1932 the firm's boat immediately became the second-leading gold producer in the state. Excellent publicity came that same year, when Spokane newspapers reported that A. A. Bennett, who had contracts to carry the mail by air between Boise and Atlanta, Cascade, and Yellow Pine, had made special flights into Warren during the winter to transport gold bullion for the firm.[23]

Idaho Gold had a few early bumps on the road to success. In late 1932 or early 1933, it had to replace the bucket line and the sluice on the boat. During the following year, for reasons that are not clear, the company's production dropped dramatically, by about 2,000 ounces. At about the same time, the firm added the *Mickey Mouse*, as it was called, a second dredge—a two-and-a-half-foot Yuba portable of all-steel construction, with pontoons that could be taken apart and moved. Originally, the small boat had been a stationary washing plant, until Yuba rebuilt it into a floating bucket dredge in 1934, first with tables, then with jigs added in the field. Company spokesmen noted that they expected the new dredge to work the upper Warren Creek until it ran out of ground, at which time it would be transported to the Elk Creek district. The little Yuba did dig almost continuously on Warren Creek from mid-July 1934 until late in the summer of 1936, when it was moved to the Boise Basin. Geologist John C. Reed, who visited Warren briefly in July 1936, estimated that in its short span the little boat had recovered roughly a quarter of a million dollars in gold and silver.[24]

After personnel of the National Geographic Society–US Geological Survey made a run down the Salmon River in the autumn of 1935, they visited the Warren area, where they saw their first bucket dredge. To them, the dredge resembled "a Mississippi River steamboat lost in the mountains." The town was booming. Three dredges and a quartz mine were working around the clock. Much ground had already been dredged, leaving behind "a gleaming, hummocky, and totally barren surface marked by crescentic bowlder ridges that cross and recross the valleys as if raked by a gigantic comb." Indeed, the little *Mickey Mouse,* so named by Al Fisher's daughter, had already passed through part of the town itself, "turning backyards into rock piles." Bulldozers had to level the lots, in the process breaking through into Mrs. McDowell's cellar where she stored all her supplies and canned goods, "like they were mining for food!"[25]

When the Idaho legislature levied a new tax on mining in 1935, Baumhoff and Fisher argued that it applied only to quartz mining and not to placer operations. Frustrated at the state's adverse interpretation, they brought suit against enforcement of the law and fought the issue through the courts, only to lose three years later when the Idaho Supreme Court handed down its decision in the case *Idaho Gold Dredging Company v. John L. Balderson.*[26]

In 1932 a second company appeared on the scene, when the Warren Creek Dredging Company built a bucket rig on Steamboat Creek southwest of town. Andy Anderson, an old hand at Idaho mining, was in charge. Joe Ferris was president of the firm and, along with Johnny Marchbank, he owned a substantial share. Ferris and Marchbank also owned a dragline dredge in the Canyon City district of Oregon on the John Day River, near the town of the same name.[27]

The small boat on Steamboat Creek was powered with electricity purchased from the Utility Gold Production Company and was capable of handling 3,500 cubic yards of gravel per day. Other details were lacking about the machine, the state inspector of mines complained, because the firm refused to file the required reports with his office. Mining engineer Corey Brayton, writing to a colleague in the summer of 1932, noted that an old Marion dredge that had been run a short time somewhere in southwestern Colorado "ten or fifteen years ago" was being rebuilt in the Warren district—the second such reconstruction in the same area in a short time.[28] In the absence of specific background information on that particular dredge, perhaps speculation as to what happened to the old Marion boat is not entirely out of the question.

Beginning in October 1932, the Warren Creek Dredging Company boat operated continuously around the clock except for minor shutdowns, although it was not large or powerful enough to dig deeply frozen ground in the coldest winter weather. At least once, operations had to be suspended temporarily when the electrical power supply was cut off when the water froze in the flume feeding the generators on the South Fork of the Salmon. Steamboat Creek was ideal for dredging. In dry seasons, the flow of water was sufficient to form a pond. The gravel averaged about twelve feet deep and contained few boulders large enough to cause trouble for a small rig. Some ground carried values of over fifty cents a yard, and the bedrock was soft and easy to clean. It is not surprising that the Warren Creek Dredging Company had a very successful and lucrative run of nearly four years. John Reed estimated that the firm took out over three-quarters of a million dollars in gold and silver under Andy Anderson's management. Anderson's death in January 1936 precipitated a crisis that led to drastic changes. At that point, Baumhoff and Fisher purchased all of the Warren Creek Company's property, including the dredge, which quickly resumed work on Steamboat Creek under its new owners in February.[29]

First, a corporate reorganization took place. The Baumhoff-Fisher Company emerged as a separate entity. So, too, did a new Warren Dredging Company, which *Minerals Yearbook* referred to as "formerly the Idaho Dredging Co." The latter did not disappear overnight, although it faded into the background. When Baumhoff and Fisher acquired new ground in 1937 and again in 1938, it was in the name of the old firm, with Fisher listed as president and Baumhoff as manager.[30]

Another change involved modernization of the old steam dredge, which was converted to electric power, the source of current the smaller Yuba had used in its day. This no doubt meant a savings in operating costs. Along with upgrading equipment, the two mining entrepreneurs were interested in expansion and were alert for new placer ground to option or buy. In early 1937 Baumhoff and his wife, E. T. Fisher, and Clara Fisher purchased the old Idaho Gold Dredging Company's Middle, North, and Canyon claims—a total of 478.9 acres—using company stock valued at $25,584. A year later Idaho Gold acquired an additional 59 acres from Andrew Kavanaugh. By this time, Baumhoff and Fisher owned all of the placer ground on Warren Creek from the town to Thomas Creek and most of the bench gravel deposits on the west side of Warren Creek Valley opposite Warren Meadows. Warren Creek above the town, Steamboat Creek, and lower Stratton and Thomas Creeks were all held under lease.[31]

The ground in all of these areas contained gold. Where the alluvial valleys were shallow and had sufficient grade for sluicing or hydraulicking, the placers had been intensively mined during the more primitive placer periods and had produced well. It was from these relatively flat grounds and from the tailings left behind by their predecessors that the dredgers found rich digging. On the Meadows especially, there was plenty of deeper virgin gravel that only the dredges could reach. The recovery was mainly coarse "shot gold" and easily caught; the small amounts of fine or flake gold could be captured by the use of quicksilver.[32]

John Reed estimated that between startup in 1931 and the end of 1936, the Idaho Gold Dredging Company's steam dredge had worked an area of roughly one-third of a square mile on Warren Meadows, with a return of between $750,000 and $1 million. A comparable amount of ground still remained to be dredged.[33] By that time the smaller machine—the old Warren Creek Dredging Company boat—was working on the lower end of Stratton Creek. Although the two dredges were on different ground, the conditions under which they operated were similar, although bedrock was deeper on the Meadows.

Some clearing was necessary in both cases, especially on the west side of the valley, where paydirt was covered with fifteen feet to twenty feet of topsoil and a thick growth of lodgepole pines. The trees were removed with a sixty horsepower Caterpillar, which left them in rows to be burned. At first, a

Cat equipped with a twelve-yard Carryall scraper was used to clear away the topsoil, but by 1938 hydraulicking had been substituted.[34]

Both dredges had the same capacity: on both, seventy-five 4-foot buckets operated at a speed of twenty-six buckets per minute, trommels were the same size, the tailings stackers were 30 feet long, and a steel spud was used for dredging and a wooden one for stepping ahead. The standard procedure in each case was to make a 5-foot cut with a 100-foot swing around the spud. Both machines averaged 90 percent operating time, although downtime in the spring when there were problems with power lines was more frequent. In addition to two Caterpillar tractors and a full complement of road-building equipment, the company owned a complete machine shop in Warren. From 1931 on it maintained its own airport, which was vital in the winter; when the company dredged the old air field, with the help of the US Forest Service and local citizens, it built a new one on the tailings.[35]

Fred Baumhoff and Al Fisher ran both boats, and they did so efficiently. Theirs was a hands-on operation, but they did not stint where staff was concerned. An experienced dredgemaster was in charge of both dredges overall. His pay was not usually made public, but the rates paid to the other crew members were. In 1937 the payroll for each boat included three winchmen, at six dollars a day each; three oilers, at five dollars a day each; one mechanic, at six dollars a day; and one shoreman, at five dollars a day. In addition, there was a general mechanic in the machine shop, an electrician, a truck driver, a Cat operator, and a bookkeeper in the office at Warren for both machines.[36]

It was calculated that overall operating expenses of the two rigs ran from 6 cents to 7 cents per cubic yard, including depreciation and depletion. Labor made up 1.37 cents of that amount; the next highest cost was replacement and repairs, at l.12 cents, followed by power, at 0.75 cents. The remainder of the cost was split among supplies, transportation, workers' compensation and unemployment insurance, and general overhead.[37]

Dredges of the two firms—the Warren Dredging Company and the Baumhoff-Fisher Company—continued to dig in the Warren district under the common leadership of Baumhoff and Fisher until early August 1939.[38] In 1941 their firm, the largest producer in the Warren district, operated two bucket dredges from January 1 to August 10, when it sold its placer ground and one dredge to W. W. Prather, who dug for the rest of the year. Prather had a good season in 1942, before the federal ban on gold mining was mandated. The two-and-a-half-foot boat recovered 1,090 fine ounces of gold and 358 ounces of silver, but no other dredge operated that year. Prather was idle in 1943, although 114 ounces of gold were cleaned up from his rig and sold. Warren's era of dredging was over.[39]

After the war, the partnership of Baumhoff and Fisher emerged as Baumhoff and Marshall, Inc. The Warren Dredging Company continued to exist, but neither company returned to digging in the Warren district. The focus now became the Boise Basin and, after that, the recovery of strategic minerals in Valley County.

Notes

1. Hiram T. French, *History of Idaho* (Chicago: Lewis, 1914), 1:29.

2. William J. Trimble, *The Mining Advance into the Inland Empire* (Madison: University of Wisconsin Press, 1914), 200, 208.

3. Waldemar Lindgren, "The Gold and Silver Veins of Silver City, Delamar, and Oro Mining Districts in Idaho," Twentieth Annual Report, part 3 (Washington, DC: US Geological Survey, 1900), 237.

4. Ibid., 240; John C. Reed, "Geology and Ore Deposits of the Warren Mining District, Idaho County, Idaho," Idaho Bureau of Mines and Geology Pamphlet no. 45 (Moscow: University of Idaho, September 1937), 4–5.

5. Trimble, *Mining Advance*, 72; Lindgren, "Gold and Silver Veins of Silver City," 238.

6. The six were J. H. Hawley, Joseph Pinkham, J. R. Richards, A. L. Richardson, E. B. True, and a man named Thompson. *Idaho Daily Statesman* (Boise), July 17, 1894.

7. Ibid., January 31, 1895; William S. Greever, *Bonanza West: The Story of the Western Mining Rushes, 1848–1900* (Boise: University of Idaho Press, 1963), 278–79.

8. *Idaho Daily Statesman* (Boise), January 31, 1895.

9. *Engineering and Mining Journal* 60 (August 3, 1895): 109.

10. *Idaho Daily Statesman* (Boise), July 6, 1897; Lindgren, "Gold and Silver Veins of Silver City," 240–41. Not much information is available on this company, which was incorporated in New Jersey with nominal capital of $150,000. *Fourteenth Annual Report of the State Board of Assessors of the State of New Jersey for the Year 1897* (Trenton, NJ: J. L. Murphy, 1898), 173.

11. *Idaho Daily Statesman* (Boise), July 6, 1897; Lindgren, "Gold and Silver Veins of Silver City," 240–41; Charles H. Munro, "Preliminary Report on Warren Meadows Dredging Venture," San Francisco, May 7, 1908, typed copy, Fred L. Morris Papers, Bancroft Library, University of California, Berkeley.

12. *Idaho Daily Statesman* (Boise), March 15, 1905; *Salt Lake Mining Review* 7 (April 30, 1905): 15; Mary Alfreda Elsensohn, *Pioneer Days in Idaho County* (Caldwell, ID: Caxton, 1957), 1:100; Munro, "Preliminary Report on Warren Meadows Dredging Venture," Fred Morris Papers.

13. Charles Munro and J. S. Wyatt to Jonathan Bourne Jr., Boise, July 13, 1907, copy; Munro to Ida Arneson, Boise, July 14, 1907, copy, both in Jonathan Bourne Jr. MSS, University of Oregon Library, Eugene.

14. J. S. Wyatt, Charles Munro, and J. S. Wallace, "Agreement," San Francisco, May 29, 1908, typed copy; Munro, "Preliminary Report on Warren Meadows Venture," San Francisco, May 7, 1908, typed copy, both in Fred Morris MSS, Bancroft Library, University of California, Berkeley. Munro was a pioneer mining engineer in the dredging field. By this time, he had successfully superintended one of the major California dredge operations for four years and was well-known as a consultant throughout the West and Alaska. *Engineering and Mining Journal* 118 (July 19, 1924): 84.

15. *Idaho Daily Statesman* (Boise), January 3, 1915.

16. Robert N. Bell, *Thirteenth Annual Report of the Mining Industry of Idaho for the Year 1911* (State Mine Inspector's Report), 69.

17. Robert N. Bell, *Twenty-First Annual Report of the Mining Industry of Idaho for the Year 1919* (State Mine Inspector's Report), 118.

18. Production statistics provided by C. N. Gerry, US Bureau of Mines, in Reed, "Geology and Ore Deposits of the Warren Mining District," 25.

19. *News-Advocate* (Price, UT), July 23, 1931; *Duchesne County (UT) Newspapers*, July 24, 1931.

20. Idaho Bureau of Mines and Geology, *Special Report* 1 (1964): 90.

21. Bruce J. Noble Jr., "A Frenchman in Wyoming: The South Pass Mining Misadventures of Emile Granier," *Wyoming Annals* (Cheyenne) (Winter 1993–94): 58; "Gold Flakes to Yellowcake Historic Mine Trail," Interpretive Plan (Cheyenne: Wyoming Historic Mine Trail and Byway Program, 2009): 7; *Minerals Yearbook, 1933* (US Geological Survey), 301; *Minerals Yearbook, 1936* (US Geological Survey), 557; *Minerals Yearbooks Review of 1940* (US Geological Survey), 479; *Minerals Yearbook, 1941* (US Geological Survey), 491; "Costs of Plant, Labor and Operating Costs," US Bureau of Mines Information Circular 6846 (Washington, DC: US Department of the Interior, Bureau of Mines, 1935), 8–9.

22. Reed, "Geology and Ore Deposits of the Warren Mining District," 44; *Mineral Resources, 1931* (US Geological Survey), 458.

23. H. S. Lorain and O. H. Metzger, "Reconnaissance of Placer Mining Districts in Idaho County, Idaho," *US Bureau of Mines Information Circular* 7023 (June 1938): 71; *Minerals Yearbook, 1932–33* (US Geological Survey), 107; *Spokesman-Review* (Spokane), April 30, 1932, clipping, Kimble Database, WSU.

24. *Minerals Yearbook, 1934* (US Geological Survey), 141–42; *Minerals Yearbook, 1935* (US Geological Survey), 207; *Minerals Yearbook, 1936*, 394, 398; *Spokesman-Review* (Spokane), August 8, 1934, clipping, Kimble Database, WSU; "List of Dredges Built by Yuba Manufacturing Company," copy, Charles Romanowitz Papers, Rasmuson Library, University of Alaska, Fairbanks; Reed, "Geology and Ore Deposits of the Warren Mining District," 45; Edward A. Foster to W. B. Macauley, June 16, 1941, Yuba Manufacturing Company MSS, Meriam Library, Chico State University, Chico, CA.

25. Philip J. Shenon and John C. Reed, "Down Idaho's River of No Return," *National Geographic Magazine* 70 (July 1936): 122, 125, 129; Cletus R. Edmunson, "A History of Warren, Idaho, Mining, Race, and Environment," MA thesis, Boise State University, Boise, ID, August 2012, 63 (all quotes).

26. *"Idaho Gold Dredging Company v. John L. Balderson," Idaho Reports* 58 (January 25, 1938): 692, 730.

27. *Minerals Yearbook, 1932–33*, 107; Reed, "Geology and Ore Deposits of the Warren Mining District," 43–44; Oregon Department of Geology and Mineral Industries, *The Ore.-Bin* (Portland: Oregon Department of Geology and Mineral Industries, February 1941), 3:13.

28. *Spokesman-Review* (Spokane), February 26, 1933, clipping, Kimble Database, WSU; Corey C. Brayton to J. E. Hopkins, August 23, 1932, copy, Alaska Gold Company MSS, University of Alaska Archives, Fairbanks.

29. Lorain and Metzger, "Reconnaissance," 76; Reed, "Geology and Ore Deposits of the Warren Mining District," 44; *Spokane Chronicle*, January 16, 1937, clipping, Kimble Database, WSU.

30. *Minerals Yearbook, 1937* (US Geological Survey), 324; *Spokane Chronicle*, January 1, 1937; January 13, 1938, clippings, Kimble Database, WSU.

31. *Spokane Chronicle*, January 1, 1937; January 13, 1938, clippings, Kimble Database, WSU; Lorain and Metzger, "Reconnaissance," 76.

32. Reed, "Geology and Ore Deposits of the Warren Mining District," 29.

33. Ibid., 45.

34. Lorain and Metzger, "Reconnaissance," 77; *Engineering and Mining Journal* 135 (February 1934): 55.

35. Lorain and Metzger, "Reconnaissance," 77–78; "Payette National Forest Warren Historic Walking Tour," 1994, http://www.secesh.net/Walk.htm.

36. Lorain and Metzger, "Reconnaissance," 78.

37. Ibid., 79.

38. *Minerals Yearbook, 1937*, 323–24; *Minerals Yearbook, 1938* (US Geological Survey), 353; *Minerals Yearbook, 1939* (US Geological Survey), 330.

39. *Minerals Yearbook, 1941*, 351; *Minerals Yearbook, 1942* (US Geological Survey), 380.

12

Boise Basin

First located in the summer of 1862, the Boise Basin would prove to be the richest and most enduring of the Idaho high-country gold regions. It was soon regarded by locals as "the greatest placer country outside of California." More gold would be produced from this area than from any other locality in the state. As usual, the first discoveries were the result of small bands of restless miners from earlier camps. In this case Moses Splawn, who had tried his hand in the placers of both Florence and Elk City, had received several tips from a friendly Bannock who had seen the precious metal in a basin farther south and east. Splawn promptly joined a party headed for the Owyhee country. The group did find some gold on Sinker Creek, but not enough to satisfy its cravings. Turning back north, the party combined with another group led by George Grimes, crossed the turbulent Snake River, and in time was able to follow the Boise River up to a canyon leading into Boise Basin. On Boston bar, near the present site of Centerville, a rich deposit of gold was discovered on August 2. While further prospecting the creek that would soon bear his name, the unfortunate Grimes was shot and killed, presumably by

DOI: 10.5876/9781607324751.c012

a hostile Boise Indian. After hastily burying Grimes in a prospect hole, the party scurried back to the Boise River and eventually to Walla Walla, where they regrouped and with reinforcements were back in the basin at fifty-two strong in early October. The word was out. The rush of eager gold seekers began immediately and peaked in 1863.[1]

Located about 300 miles southeast of Walla Walla and seventy miles due east of the old Fort Boise, Boise Basin attracted the overflow not only from earlier mining camps but from neighboring states as well. Many Californians found their way overland across Nevada or they came by sea, then up the Columbia River by steamboat to Umatilla Landing, the head of navigation—which left another 300 miles over terrain far less difficult than that leading to other Idaho high placer fields. Access to the basin itself was much easier, too. Near the end of the Civil War, a good number of "the left wing of Price's army" arrived from southern and border states, which stirred rigorous partisan friction.[2] Ultimately, completion of the transcontinental railroad in 1869 provided an easier route both from the east and from California. The distance from Kelton, just north of Salt Lake City, to the Boise Basin was about 40 miles shorter than that from Walla Walla or Umatilla, and it quickly became the preferred road from the east for both freight and passenger service. Completion of the Oregon Short Line railroad in 1884 put Kelton out of business and greatly facilitated transportation to southwest Idaho.[3]

Not only was it richer in precious metals, but the Boise Basin took on an aura of permanency much earlier and to a greater degree than any of the other high placers. While by no means salubrious, the climate in the basin was more temperate than that in the other high camps, which gave its settlements more year-round stability and encouraged miners to bring their families earlier. Proximity to the much-traveled Oregon Trail helped increase the numbers of men, women, and children on the move in the 1860s. The location of the town of Boise, twenty-five miles away, along with a military post—both established in 1863—created a rapidly growing important trading center through which passed much of the basin's business. The fertile valleys of the Payette River just to the north and the Boise River to the south were quickly taken up by settlers who were able to supply many of the needs of the mining communities, as well as those of the surrounding area.[4]

In his 1896 geological survey of the Boise Basin, Waldemar Lindgren consistently referred to that area as the Idaho Basin. Lindgren noted that the productive part of the basin measures 15 miles in length from north to south and in maximum width approximately 13 miles—a total of not over 150 square miles. Later geologists enlarged this by 100–150 or so square miles, although Hubert Howe Bancroft called it a "picturesque depression among the mountains about thirty miles square," and George Crofutt's 1870 *Great Trans-Continental Tourist's Guide* listed its dimensions as a whopping 150 miles north to south by a length of about 200 miles.[5] Boise Basin was not square, as Bancroft implied; it was more fan-shaped. Like the other Idaho high-country placer areas, it was certainly scenic. Elijah Chaney, who visited there in the summer of 1866, was enchanted by the setting, "embraced within a perfect circle, formed by the higher ranges of the Blue Mountains that bound it." "It was a sublime scene of wondrous beauty," he wrote. On the grand ridge were 100 snow-capped crests, "each dotted with pine sentinels and blue groves." Below was a second concentric ring of "ninety low mountains, nearly of an equal height . . . as beautiful as the heaving waves of the stormy ocean." Along with the Blue Mountains were the high divides of the Boise Mountains and irregular spurs of the Sawtooth range. A number of prominent peaks surrounding the basin reached more than 8,000 feet in altitude, although most of the towns and the placer areas, except Summit Flat, were slightly more than half that high.[6]

In this environment, towns sprang up overnight, beginning in late 1862. The basin's original production of gold was from the rich creek placers, but lode mining quickly followed. Although there were exceptions like the Gold Hill mine near Quartzburg, which was worked almost continuously from its discovery in 1864 until 1938, lode mining tended to be much more erratic. Ore bodies gave out, and often deeper sulfide ores required more complex treatment than the old-fashioned stamp-amalgamation process. In contrast, the free gold of the placers sustained a high output until close to the end of the nineteenth century. The most productive placer portion of the basin was its central and southern part, with focus mainly on Idaho City, Mores Creek, Gambrinus, and Centerville. The Pioneerville district, which included Summit Flat and Grimes Pass as well as the town of Pioneerville, was in the northern part of Boise Basin and was primarily a lode mining area, in which gold production was heaviest prior to 1920. The Quartzburg district, which included the town of the same name, along with Gold Hill, Granite, and Placerville, supported both kinds of mining, although there, too, quartz was most prominent.[7]

In the basin itself, the flat bottomlands were separated into two valleys by Grimes Creek and Mores Creek, each drained by one of the streams, with a high divide between. On the hillsides above the towns, hydraulic mining flourished. In 1866 at Idaho City, Elijah Chaney commented on the "many hydraulic pipes" at work on the slopes higher up—"the mournful roaring of them mingling together, making music pleasing to hear." Leaving the basin, both watercourses flowed for several miles through deep canyons before uniting to form Mores Creek, which eventually emptied into the Boise River.[8]

One characteristic of the Boise Basin that set it apart from the other high placer mining regions was the multiplicity of towns that cropped up in that relatively small area. In Pierce, Florence, Elk City, Warren, and the other regions, only one or two major settlements usually existed. Idaho City was the largest in the basin and became the county seat, but half a dozen other local centers were not insignificant in the early years. According to the 1863 Territorial Census, Idaho City had a population of 6,275, Placerville counted 3,254, Pioneer City 2,743, and Centerville 2,638. At the peak of the rush a year later, Idaho City's population had reached 7,000—the largest in the Northwest—but after the rich easy-picking mining had fallen off, in 1870 it had dropped to a mere 889, half of which were Chinese. In addition, many now-forgotten smaller settlements cropped up from time to time, enjoyed their brief moment in history, then passed into obscurity—Pine Grove, Pomona, Boston, Moorstown, and Granite City among them. Others, like Buena Vista—across Elk Creek from Idaho City—were victims of dredging, which disfigured a good deal of low-lying land in the basin. Idaho City was leveled by fire several times in the 1860s but quickly rebuilt. Flames consumed most of Quartzburg in 1931, and the town never rose from the ashes.[9]

Widely circulated production estimates for the basin's first thirty-five years ran as high as $100 million. While agreeing that this relatively small area had produced a disproportionate amount of gold, most of it from placers, Waldemar Lindgren rejected that figure: "Allowing for the uncertainty in the data, we may safely say that the basin had not produced more than $50,000,000." Of this, recovery from quartz mines in the same 1863 to 1898 period accounted for $4 million.[10]

Precisely when the first dredge appeared on the Boise Basin scene is not clear. According to Idaho historian Arthur Hart, the Bucyrus Company sent an agent there in 1893 trying to sell its "gold-saving machine," and one such dredge was actually working at that time, but there is no further mention of this equipment as such. Lindgren reported that a dredge had operated in 1896 on lower Mores Creek, about four-and-a-half miles above the mouth of Grimes Creek, and it succeeded in reaching rich gravel between 10 feet and 20 feet down, despite large boulders that provided obstacles to easy digging. This was relatively narrow ground covered with tailings washed down the hills by spring and winter freshets and by Chinese miners using dams, Chinese pumps, and derricks. Lindgren gives no clue as to what kind of dredge it was or if it was continuing to work, but he did note that lower Mores Creek gravels were not "considered to have been extremely rich." On the upper part of the same stream, near Idaho City, tailings were more valuable, and they widened to a maximum of 2,000 feet and were as deep as 40 feet. Here, Lindgren believed the only means of profitable gold recovery was by dredges or hydraulic elevators.[11]

Within a year or two, the success of bucket-line dredges elsewhere had a significant impact on Boise Basin. First, there was a scramble for good dredge ground, which the basin had in abundance; then, dredge boats began to pop up like mushrooms.

In the scuffle for suitable dredging property at the turn of the twentieth century, Boise Basin loomed large. Both eastern and western capitalists were competing for a slice of the rich pie. The Basic Company was incorporated in New Jersey in 1898 by four Philadelphians, of whom Herman Hoopes and Howard W. Lippincott were probably the most important. A graduate of Swarthmore College, Hoopes wore several hats: once a partner in a paper mill, also a banker, stock manipulator, and gentleman farmer. Lippincott was a broker and a "member of one of Philadelphia's oldest Quaker family [*sic*]." Both he and Lippincott served on the Board of Managers of Swarthmore College, along with Joseph Wharton, one of the founders of the school. A highly successful industrialist in zinc, nickel, copper, and iron, Wharton was also responsible for the establishment of the University of Pennsylvania's School of Finance and Economy.[12] All three men—Hoopes, Lippincott, and Wharton—would be involved in Boise Basin gold-dredging concerns spun off by the Basic Company, which acted essentially as an overall umbrella.

The Basic Company created four subsidiaries, all chartered in West Virginia and all with very little paid-in capital. One, the Bed Rock Dredging Company, nominally capitalized at $1 million in shares of $100 each, was the first, incorporated on July 19, 1897. Incorporators were Herman Hoopes, John H. Shrite, J. W. Bethell, and Henry G. Hart, all of Philadelphia, along with Samuel S. Harper of Denver. Another, the Boise Dredging Company, chartered on February 12, 1898, also had nominal capital of $1 million. Incorporators included Hoopes and Harper plus Howard W. Lippincott and Thomas B. Turner of Philadelphia, as well as Edwin A. Barber of Westchester, Pennsylvania.[13] A third charter, Union Placer Dredging and Refining, appeared in March 1898. Its authorized capital was listed as $600,000. The company had $2,000 paid-in and a completely different group of charterers: J. A. McDuffie of New York City and Samuel F. Adams, Charles H. Roney, Charles C. Warwick, and George McClean, all of Philadelphia. Finally, dated January 18, 1899, the Bullion Dredging Company was created, which had the usual $1 million capitalization and the familiar incorporators Hoopes, Lippincott, Barber, Harper, and Bethell. Archaeologist Edwin Barber, who had signed the charters of two of the corporations, must have been very proud of his venture into mining. In his biographical sketch in *Who's Who in America,* he included not only a reference to his book, *History of Tobacco Pipe,* but he added the words "engaged in gold dredging."[14]

In early 1898 the Idaho City press indicated that the Montana Dredge Company had purchased ground at Placerville, bonded ground on Grimes Creek from S. K. Goldtrap, and expected to employ 100 to 150 men during the coming summer. Wages were to be three dollars a day, and the workers would be allowed to board where they pleased when working near town; when farther away, the company would run a boardinghouse, with meals furnished at actual cost. The Boise paper indicated that testing on the Boyle Gulch property, bonded by "Philadelphia parties," had struck bedrock at about thirty-five feet, whereupon the company had made the owners a cash proposition to buy the ground at once. The same group had also taken a bond on the Goldtrap and Bowen Placers that extended five miles along Grimes Creek and now moved to prospect that ground at a point about a mile above Centerville.[15]

Already, more men were employed than at any previous time. At Placerville, a large force was constructing a boat and doing other work for the Montana Placer Dredge Company, while more than 100 others were digging roads and erecting buildings four miles below Centerville. (The Montana Placer Dredge Company seems to have been the Boise Dredging Company.)

FIGURE 12.1. One of the first dredges in Boise Basin, the Bed Rock machine was a failure, but in time, with modification, it achieved some success. Courtesy, Idaho State Historical Society, Boise, ISHS 72-112.b.

Moreover, the press carried the news that the companies intended to increase the Centerville force to 150 as soon as structures for their accommodation were ready. A day later, a spokesman for the Basic and Bed Rock Companies was refuting these words and advertising that they were no longer looking for laborers. Men were warned not to expect work if they came to the basin.[16] The general manager of the Basic Company was Samuel S. Harper, the man primarily responsible for designing, overseeing construction, and promoting the first successful gold dredge in Montana. Harper was no doubt involved in organizing the Basic Company. As general manager, he employed Theodore F. Van Wagenen, a respectable Colorado engineer, as superintendent. In that capacity, Van Wagenen was given extensive power to look after the affairs of the sub-companies, including further land acquisition on their behalf.[17] The Basic Company itself controlled twenty-four miles of bottoms along Grimes,

Wolf, and Granite Creeks and in Bayles's Gulch—enough placer ground, its spokesman believed, to keep eight to ten modern bucket-line dredges busy for some time.[18]

One of Basic's first charges was to install a hydroelectric plant on Grimes Creek to provide electric current for the Bed Rock Dredging Company at Placerville and the Boise Dredging Company at Centerville, as well as for other possible users. Samuel S. Harper's pioneering dredge at Bannack had been electric, not steam-powered. The task was not easy. First, it was necessary to cut a shelf 7 feet wide out of the side of a mountain. Next was the construction of an 8½-mile combined ditch and flume, by which to divert the stream. Of that length, almost 3 miles were ditch and 5.3 miles were wooden flume, with one 163-foot tunnel. Approximately 766,900 board-feet of lumber and 21,700 pounds of nails were used. The work took 121 days, but the canal route was so uniformly steep that was possible to utilize one team with a plow and scraper for only 16½ days. The remainder of the excavation was by pick, shovel, and drill performed by 175 laborers, directed by foreman Alexander Kerr and flume builder Chase Wilson. From a pressure box at the end of the canal, the water was dropped 350 feet through a 26-inch steel pipe, which was gradually reduced to 15 inches, from which it was led through a cast-iron nozzle piece in two jets to the Pelton wheel in the powerhouse. By early August 1898, wires were up between the corrugated iron powerhouse and the Placerville dredge, a distance of 15 miles; a telephone system had been installed; and Superintendent Van Wagenen was certain that the boat at Centerville would be finished in time for a fall run of a month or so. The company expected the last piece of its electrical equipment, a nearly 10-ton cast foundation for the dynamo, to arrive by the end of August. The firm was so concerned about the machine's weight on poor roads that it sent east for an experienced "rigger," a specialist in moving heavy machinery. Even then, the arrival was later than expected. With everything finally in place and a Westinghouse two-phase generator connected directly to the Pelton water wheel, the Bed Rock Dredging Company's Bucyrus-built boat was the first gold dredge in the country to use alternating current and poly-phase motors for its work. Bearing the name *H. W. Lippincott,* the big machine started digging on September 28.[19]

The Boise Dredging Company completed its dredge half a mile above Centerville soon thereafter. Its machine was named the *S. K. Goldtrap*, after the local owner of placer ground, and was a copy of the Placerville boat, both designed by Sam Harper. Another boat was built for the Union Placer Dredging and Refining Company, a mile below Centerville on Golden Gulch bar. The hull was put up in 1898, but it was not completed until 1902. The Bullion Dredging Company also had a boat under construction. All of these dredges were alike, although only the Bed Rock machine was manufactured by Bucyrus. All of them were electrically driven, had five-foot open-connected buckets, and were of the double-lift type, with a boulder chute and a scow behind to carry the long sluice.[20]

With this early beginning, the future of dredge mining in the area looked bright indeed. In the summer of 1899, *Mining and Scientific Press* made an optimistic forecast: "It is not impossible, nor improbable, that Boise Basin will contain twenty-five operating dredgers within three years."[21]

Such was not the case. The staying power of the early dredges was not great, and those of the Basic Company's subordinate firms were no exception. The Bed Rock Company's *H. W. Lippincott* finished its first brief season and worked for a month before shutting down. In the autumn of 1900, it was rebuilt. In early 1904 the boat was for sale, "practically good as new." The Bullion Dredging Company's boat had not yet been completed in late October 1900; when finished, it operated for only one season on Grimes Creek. The Boise Dredging Company's *S. K. Goldtrap* dug just a short time before it was idled and remained so for two years. The dredge of the Centerville Gold Dredging Company operated the entire season in 1899, was remodeled in 1900, worked all of the 1901 season, and ran only two months in 1902 and part of the 1903 season, after which it, too, closed down.[22] When he visited Boise Basin in early 1899, J.U.O. Bethel, secretary of the Bed Rock Company, announced that his concern would put another dredge on its ground at once, but that did not happen. In November, at the close of the season when the boat had shut down for the winter, he was lamenting the poor quality of the firm's machinery. At Placerville, the *H. W. Lippincott* had lost "about seven-tenths" of the work year because of breakdowns. At Centerville, certain parts of the *S. K. Goldtrap* had given out entirely, and the plant was in need of a thorough overhaul. According to a veteran mining engineer in 1902, much of the equipment purchased by the companies "was of very inferior quality and broke faster than it could be repaired." Experts charged that the $300,000 power plant on Grimes Creek was inadequate and poorly constructed. Overall, the dredges produced only about 60 percent of sample values.[23]

In addition, the companies were on shaky economic ground from the beginning. Even before the dredges and electric power plant were completed, their boards of trustees mortgaged the property, with much of the loan money provided by the industrial giant Joseph Wharton. When coupled with the lack of success in actual dredging, "either because of incompetency of those in charge or of flaws in the castings; perhaps both," receivership was sometimes unavoidable. This was the case with the Boise Dredging Company in May 1901, when a decree of foreclosure and public sale of all property were announced—including the placer ground, the building, the dredge *S. K. Goldtrap,* "and all the machinery, equipment, belting, shovels, buckets, guide ropes, poles, wire, electric equipment and accessories of every sort connected therewith." The boat was purchased by Charles W. Burr at the bargain price of $475 and subsequently resold to the Centerville Dredging Company, a new concern chartered in New Jersey. Although it was reconstructed several times, there is little evidence that the dredge ever turned a profit.[24]

Thus at Centerville, in reorganization, the holdings of the old Boise Dredging Company were divided. The dredge below town, the power plant, and part of the land passed into new hands. A fresh manager, T. G. Janney, would oversee some of the changes made on the machine, which then began operations in the spring season of 1902. Theoretically, the boat was capable of handling 3,500 yards per twenty-four hours; in actuality, during its 1903 run it managed to work only 55 percent of that amount.[25]

The parent firm, the Basic Company, had troubles of its own. When its superintendent, Theodore Van Wagenen, resigned sometime in early 1899, the firm brought suit against him, alleging that while employed he had violated an agreement not to acquire placer ground for his own use. Instead, he had bonded property owned by S. K. Goldtrap with the intent to prospect and, if it proved valuable, to put a dredge on it. Van Wagenen proved that he had taken the option after he had left the company and won his case.[26]

This was a minor affair, however. The next event would be much more serious. In 1902 Norman Gratz came to the basin from Philadelphia as manager of the Union Dredging Company. He seems to have been selected by the influential power behind the scenes, Joseph Wharton. Young Gratz, who was barely twenty-four, had majored in chemical engineering at the University of Pennsylvania but had not completed his studies because of poor health. Whether the boat was digging that year is not certain, but in 1903 it started up in April and was working three shifts in satisfactory fashion. Its power came from the Grimes Creek power plant, and the necessary water was obtained from that stream and from Granite and Wolf Creeks. Apparently, there was a smaller dredge powered by the same plant but operated by a sub-company. At the end of 1904, Gratz announced that the Union Dredging Company's large machine was in "thoroughly good condition preparatory to a long run next year." If the season proved successful, a second dredge might be installed. Of even more interest was his statement that a new power plant would be built on the South Fork of the Payette, mainly with the idea of producing electricity primarily for the use of others. Gratz's company would continue to use the old Grimes Creek plant.[27] Construction of the dam was nearly complete in mid-March 1907, but the installation of machinery would take several more months. Unfortunately for Gratz, when finished, the Wharton dam, as it was called, broke before the power line could be strung. Gratz went east in July 1908 to try to induce men of means to rebuild. Rebuilding did occur, but only after the old site had been acquired by the Boston and Idaho Gold Dredging Company.[28]

Meanwhile, both the Basic Company and the Union Dredging Company were going through troubled times. Joseph Wharton's biographer tells the story of how in 1904, Wharton sought to take over Union Dredging using a technique he had found useful on prior occasions to gain control of foundering corporations: buy the company's stock at a bargain price, force it into bankruptcy, then purchase its property at a US marshal's sale. In this case, he successfully acquired almost all of Union Gold's worthless stock, but one shareholder, Herman Hoopes, and his employer, the Westinghouse Company, combined to halt the bankruptcy in court, leading to several years of litigation. Meanwhile, the company was permitted to continue its preparations to begin dredging, but Wharton faced the same expensive obstacles that had shackled Union Gold earlier: building access roads and constructing a dam and power station in 1906–7 on the Payette River—which were expensive and time-consuming. When the General Electric Company missed its deadline for equipment and sent it so late that it delayed the project for a year, Wharton refused to pay, and more litigation ensued. When Wharton died in early 1909, the dam was still unfinished.[29]

Before his passing, Wharton called this Idaho episode "the worst defeat I have ever suffered" and blamed two people for the fiasco. One was his hand-picked manager, Norman Gratz, who was at the helm for seven years in Idaho after his appointment. The other was a young lawyer named William E. Borah, whose legal work on the firm's behalf was "highly stupid and unsatisfactory" and had resulted in failure to prevent the injunction barring Wharton from having the company declared insolvent. "Borah seems to have thought that I was a stupid old Quaker without any influence, whom he could injure without any harm to himself," Wharton wrote, vowing to inform the freshman US senator from Idaho's new colleagues about just what kind of man he was.[30]

This may have happened, but the account told by the press and the court records has a different twist. The case of *Herman Hoopes et al. v. the Basic Company* dragged on for nearly three years before it was settled. Its proceedings revealed a great deal about corporate enterprise in Boise Basin. When Hoopes, Howard Lippincott, and others organized the Basic Company, they spent most of their capital on securing title to placer ground. Since subsequent operations were not very successful, to meet the need for more cash to pay for additional property, water rights, and machinery, the firm borrowed money from Joseph Wharton. The millionaire industrialist advanced a total of $110,000, with two separate mortgages, taking as security the holdings of the Basic Company. Later, the Union Dredging Company took over Basic's holdings; still later, Union Dredging settled full claims by winding up its affairs and conveying all of its holdings to Wharton. This action brought charges of fraudulent collusion by Wharton and the majority of shareholders of Union Dredging from Herman Hoopes, who secured a temporary injunction blocking the transfer. Hoopes lost the lawsuit when it became known that he no longer held any stock in the Basic Company: before he sued, he had transferred all of his shares in that concern to the Union Dredging Company in exchange for $60,000 par value of stock in that firm. Judge Wood ruled that there had been no hint of fraud and that Joseph Wharton was the sole owner of all property heretofore held by the two companies.[31] Wharton died in early 1909, and the basin property in his estate sat idle for a number of years.

Not every early Boise Basin dredging company went through the financial agonies of the Philadelphia group. The Bostonians fared much better. In autumn of 1897 there was a fierce bidding contest for property known as the

"Idaho City bedrock flume," a little over a thousand acres that ran seven miles down Mores Creek from Idaho City. On one side were "some of Boston's big guns," headed by Charles H. Souther, as opposed to a Montana coalition, mainly from the dredge mining operation at Bannack, whose spokesman was M. F. Kirkpatrick. Souther won with a bid of $36,000, which was accepted before word of Kirkpatrick's higher offer reached the sellers' agent. An unhappy Kirkpatrick cried "foul" but perhaps found some consolation in having purchased between 400 acres and 500 acres of placer ground at Placerville earlier.[32]

Charles H. Souther was the son of John Souther, a very prominent heavy equipment builder who invented the original single-bucket dredge and established Globe Locomotive Works in south Boston. The senior Souther's dredging machines became "known the world over as the best." Young Charles went into business with his father at age twenty-three in 1873, when he became president of the New England Dredging Company. Until early in the twentieth century, this firm specialized in clearing Massachusetts harbors and channels for federal, state, and city governments.[33]

Seeking further investment opportunities, Charles Souther had California landholdings by the early 1890s and must have been aware of the many failed gold-dredging attempts in that state's rivers.[34] Now that the technology had proven itself, he was one of the leaders in applying it in the Boise Basin.

When the younger Souther first took the option on the Idaho City property, it was in the name of the New England Dredging Company. Charles W. Gardner was hired to do the prospecting and set out to begin the work on the ground in November 1897. He soon encountered a problem—namely, that no sooner had he sunk a shaft than water filled the hole. Souther then recalled that he had seen a Keystone drill at a dam site in Massachusetts and bought a No. 2 traction drill, equipped with a vacuum and sand pump, which was shipped to Idaho City and took care of the water difficulties. Souther liked the drill so much that he purchased a second one for his California property. Others followed suit, and the Keystone equipment soon became standard for placer mining.[35]

Once the testing had shown profitable values, the Boston and Idaho Gold Dredging Company was organized to work the ground. Capital came from both Boston and Idaho. Charles and John Souther were involved, with a controlling interest. Charles W. Gardner, Captain Rupert Winters, and William

S. Crosby were also investors, the latter a South Boston banker and one of the old Hawes School boys, along with both Charles and John Souther. Idaho shareholders included Governor Frank Steunenberg and future governor James H. Hawley. Gardner was named general manager and Winters superintendent in charge of the dredge.[36]

From Risdon Iron Works, at a cost of either $38,000 or $40,000,[37] Charles Souther ordered a 3¼-foot steam-powered, wood-hulled dredge with an open-connected bucket line—the seventh such machine built by the San Francisco firm. In late April 1898 Risdon's representative, D. Cameron, arrived to take charge of erecting the new boat. He came directly from Breckenridge, Colorado, where he had just supervised the construction of two dredges in deep snow. With big shipments of heavy equipment expected to arrive by rail in Boise in a few days, Cameron immediately put out a call for bids to haul 200 tons of freight from the Boise depot to Warm Springs near Idaho City. He apologized for the delay in San Francisco and explained that, because of the Spanish-American War, every US ship coming into the port had to be overhauled—a task "which has taken every mechanic in the city." Even so, he expected the machinery to arrive by the time the hull was ready and the completed dredge to be ready to run between June 25 and July 1. He also took the occasion to give the kind of pep talk that might be expected of a dredge manufacturer. The basin was ideal for that kind of mining, Cameron said. The ground was rich, and there were vast flat areas to be worked. There was room for at least a dozen boats. While Cameron was still on the scene, Robert H. Postlethwaite, the transplanted New Zealander who was the general manager of Risdon's dredge department, stopped by to add his encouragement.[38]

Precisely when the Boston and Idaho Gold Dredging Company's dredge started up is not recorded; certainly, it was not on the schedule Cameron envisioned. By the end of August 1898, the dredge was working and a night shift was to be put on the next week, thanks to Henry Schmelzel, who came to the basin earlier from Boise to install the electric light plant at Warm Springs. Later, the crews were shifted to three eight-hour shifts.[39] The little Risdon could handle from 1,500 yards to 2,500 yards of gravel per day at an expense of 2 cents a yard, with an average yield of $240 per diem, according to "a reliable source"—one that grossly exaggerated the machine's capacity. The corporate officers were less open. For example, in December at the end of the first season, when the boat closed down because of the weather, Charles Gardner was interviewed by a *Statesman* reporter as he headed "home" to Boston. His "olfactories long for a whiff of baked beans and brown bread," the newsman said, but when pumped for specific information about values of the gold and how much had been recovered, Gardner responded "with true Bostonian politeness," yet he would give no detail except to say he was pleased with the season's work. When asked about the average gold content per yard, the manager's response was "it cost us about $15,000 to ascertain that." Charles Souther was equally reticent about speaking of the property in terms of dollars and cents.[40]

Souther did express some dissatisfaction with the Risdon and noted that the company contemplated either a new dredge or extensive improvements on the old machine. He seemed to believe that one large shovel or scoop bucket might be more effective than an endless chain of buckets, which caused a continuous commotion that, in turn, contributed to a loss of precious metal. Accordingly, in late April 1899 he dispatched a corporate envoy, Dudley Dean, to Portland and San Francisco, taking plans to be submitted to dredge manufacturers in those cities. Another Bostonian, Dean was an All-American quarterback from Harvard, as well as a hero from the dash up San Juan Hill with Teddy Roosevelt.[41]

Five months later, superintendent Rupert Winters announced that machinery was being placed in the new dredge, and it was expected to start up by October 1. The old Risdon was still working steadily, but it only handled from 1,000 yards to 1,200 yards a day. Winters would give no particulars on returns, but the usual "reliable sources" reported that recovery surpassed the prospecting figures of an average of 42 cents a yard, which would make the gross daily output well over $400. The word was that the Risdon had paid for itself and the property and that its production in the first three years was around $150,000. The old boat was run with seven men—four on the daily shift and three at night—with daily operating expenses, including salaries, of about $50, of which $9 was for three cords of wood for fuel.[42]

The company's second dredge was designed by engineer Joel Brown and went into action in the fall of 1900, with W. N. Nostrand in charge. It employed a single-scoop two-yard bucket, with a long boom to dump into a hopper once every minute. From there it was "hydraulicked" to a washing plant equipped with two grizzlies. For a twenty-four-hour day, it required eight men and ten cords of wood. At a cost of $80,000, it was twice as expensive as the old rig, but it was expected to have twice the capacity. Both machines continued to work for a time on Mores Creek and its confluents until 1905. According to the *Engineering and Mining Journal*, in 1904, its sixth season, the little Risdon was usually digging down about twenty-five feet to bedrock, where its gold content averaged approximately 20 cents a yard and the operating cost between 6 cents and 7 cents a yard. The dipper dredge turned out to be a costly mistake. It handled a vast amount of gravel but was unable to save enough gold to make it a success. In addition, it constantly needed repairs. Although it was remodeled several times, by late 1903 the company decided that dismantling the dredge was the answer, but when that occurred is uncertain.[43]

When the scoop dredge was under construction in 1899, Winters talked openly of adding two additional dredges in the basin—one at Cold Springs and the other at the junction of Mores and Elk Creeks—but he made it clear that no final decision had been. Under Charles Souther's direction, the Boston and Idaho Company had also sought dredging ground outside Boise Basin. Acting as financial manager of a new California mining firm in April 1899, Souther himself bonded a huge block of property on Yreka Creek running from the Shasta River almost to the Yreka town site boundary. A few months later Winters indicated that the firm had a party of prospectors in a large tract of land in Stanley Basin. That may have been the ground the company still owned on Yankee Fork nearly a decade later, when there was talk about putting a dredge to work there. In 1905 the Boston and Idaho people also had an interest in Junction bar property on the South Fork of the Boise River.[44] The company followed through on the California enterprise but did not develop any of the Idaho possibilities.

As majority stockholders, the Southers had strong input in shaping policies of the board of directors in Boston, but they spent little time at the dredge. Charles might visit occasionally for a week or so at a time, but John apparently did not. General manager Charles Gardner made frequent inspection trips, but superintendent Rupert Winters was practically in charge and was responsible for carrying out the board's decisions and smoothing out day-to-day operations. Near the end of 1901, Gardner was named general manager of the Squaw Creek Mining Company in California, a corporation organized by the same people who operated Boston and Idaho dredges. At the same time, Winters took over the vacant general manager-ship in the basin and continued to act as superintendent.[45]

In early 1902 Winters pushed to electrify both steam dredges as an economy measure, and the company applied to the Boise–Payette River Electric Company for power.[46] The issue was not resolved and was still alive a year-and-a-half later when the company was again considering increasing the number of dredges. Adding more machinery was a question on which corporate members failed to concur, but eventually they did agree to hire a consultant to examine their property to determine if such a move were actually warranted.[47] A mining expert, Forbes Rickard, was brought in from Colorado, where he was practicing. Rickard came from a large extended family of British engineers scattered around the globe.

After a week in Boise Basin, Rickard came up with several conclusions, the first of which was that the company's ground was indeed extensive and rich enough to justify at least two and possibly three new dredges—one at Cold Springs, six miles below Idaho City; another below the junction of Mores and Elk Creeks; and if there were a third, it should be on Mores Creek a short distance above the junction. He was adamant that the new boats be electrically powered. If it could not buy the required power, the company should build its own electric plant, if need be, using the groves of timber on its land as fuel to produce steam with which to generate electricity. Rickard scoffed at the inefficiency of the scoop dredge and noted that the Risdon, which could handle 1,000 yards a day when new, was now able to take care of only 850 yards each day. The new machines should be of the Bucyrus bucket-line type, capable of digging 2,500 yards in a twenty-four-hour day, with the Robbins improved stacker. Prices on the machines would range from $60,000 to $70,000 each, he said, depending on the cost of iron and steel at the time of manufacture. Winters agreed with Rickard, but little was done to carry out the recommendations, except to remodel the dipper dredge again to put it back in operation.[48] Indeed, it would be at least five years before changes were made.

Charles Souther died in early 1905 at age fifty-five. His father, John, survived him by six years, then expired when he was ninety-five.[49] After a trip to the "hub" city to confer with officers of the firm, Rupert Winters reported back to his cohorts in the basin that all of the complications growing out of Souther's demise had been resolved and that the way was now clear to proceed with the purchase of two new dredges. The fate of the little Risdon is unsettled. According to one source, in anticipation of acquiring new equipment, the Boston and Idaho Company did not operate its small boat in 1905, in the belief that it would be cheaper to let it remain idle and handle the ground with larger plants. John H. Miles, who sometime before 1912 had replaced Winters as superintendent, wrote that the little dredge sank in the fall of 1905 and was deemed too old to raise. Unfortunately, the two leading manufacturers of such equipment, unnamed but probably Yuba Manufacturing and Bucyrus, already had orders for six dredges, commitments that would tie up their plants for many months. Then there was the old bugaboo of securing electrical power. For the new machines, Winters estimated that 500 horsepower would be required, but no power plant was available that was willing to sign a contract to deliver it.[50] Whether Winters was dealing with the old board of directors or a new one is not explained. What seems clear is that no actual dredging was done for several years.

With the passing of Charles Souther, control of the company shifted to Winthrop W. Estabrook, another Bostonian. Estabrook was young. He graduated from Harvard in 1905 but already had mining interests in California, and he had the wherewithal to take over a majority of the stock.[51] According to the press, Eastabrook did not take charge of the Boston and Idaho Company until 1907. The same reporter observed that the arrival of the new president-manager in Idaho City was met with some skepticism by the local populace—"for here was not only a man who was so unwise as to undertake to 'pan' gold by machinery, but who also smoked Turkish cigarettes, essayed Messiah-like whiskers and rode over the mountains in an automobile!"[52]

Unlike his predecessor, Estabrook was an excellent administrator who immediately made his presence felt. He moved to Boise, established an office, and promptly began a systematic program of test drilling the company's placer ground. He found values varying from 16 cents to 18 cents a yard, with fineness averaging 780, adding up to $16.54 an ounce. Capital stock of the Boston and Idaho Gold Dredging Company was increased in 1906 from $500,000 to $1 million.[53] Big changes were in the wind, but they were not apparent until one Wednesday in late September 1908, when half of a thirty-five-foot-long bucket ladder drawn by a twelve-horse team arrived from the rail head in Boise. It was the first part of a five-foot Risdon dredge sent from Yreka, California, to be installed about six miles below Idaho City at Cold

Springs. Heated by steam so it could run all winter, the machine was electrically powered. The company had rebuilt the old Wharton dam, at a considerable cost that included the death of employee William McFarland, who was killed when timbers came down a lumber chute and struck him in the head. A lawsuit brought by his wife, charging carelessness and negligence on the part of the Boston and Idaho firm, was subsequently thrown out, as usually happened in cases of this sort in that early period. The fifty-foot rock-filled timber dam diverted the South Fork of the Payette River at a low stage six miles above Garden Valley, where the company built a 1,400 horsepower hydroelectric power plant. Two horizontal Leffel turbines, with Lombard governors, were directly connected with two 450 kilowatt generators to supply current sent fifteen miles to the dredges and to the towns of Idaho City, Centerville, Placerville, and Quartzburg. Both dredge and power plant were in operation by the spring of 1909. Working three shifts, this boat proved quite successful, especially after its open-connected buckets were changed to the close-connected type.[54]

At the same time, the company let it be known that it was having a monster dredge designed that would be built by the Yuba Manufacturing Company, to have a daily capacity of 8,000 cubic yards—four times as much as the 5-foot Risdon. Actually, this was another hybrid machine—designed by the Bucyrus Company, bought by Yuba Manufacturing, then modified and erected for the Boston and Idaho firm. The Bucyrus records show it as having 13½- foot buckets; Yuba Manufacturing lists it with 15-foot buckets. During its lifetime, the dredge used buckets of both capacities plus 18- and 21-footers, although the latter did not work well.[55] It was expected to cost "in the neighborhood of $140,000." When completed, the price tag was said to have been $240,000.[56] Installation would begin on about July 1, 1910, at the junction of Elk Creek and North Creek, roughly a mile north of Idaho City.[57]

In May of the following year, the big machine made its maiden run. After a brief adjustment period it settled into a smooth-running routine,[58] an event that signified that the Boston and Idaho Gold Dredging Company was no longer a bit player on the Boise Basin mining scene. By Idaho dredging standards, the new boat was truly immense, although in the California fields there were half-a-dozen as large. The hull was built of 700,000 board-feet of Ponderosa pine, cut from the company's property and sawed in the company sawmill. The digging ladder was 97 feet long and weighed 60 tons. Each of

FIGURE 12.2. This unidentified photo showing buckets for either a new dredge or a new bucket line has all the earmarks of the Boston and Idaho Gold Dredging Company's huge new machine, which started up in 1911. Courtesy, Idaho State Historical Society, Boise, ISHS 72-112.b.

its sixty-eight 15-foot buckets weighed 4,200 pounds, including the pin. Its grizzly was 50.5 feet long and 9 feet in diameter, and it weighed almost 56 tons. The total weight of machinery on the dredge was said to be roughly 900 tons. High railroad freight rates plus even higher charges for hauling such heavy equipment between Boise and the worksite with twenty-horse teams had to be a challenge, both economically and physically. According to the state inspector of mines in 1910, the company had about 25 million cubic yards of pay gravel available, which averaged around 25 feet deep and included an overburden of from 6 feet to 12 feet of tailings carried down from earlier placer operations. An imaginative *Statesman* reporter called Souther's little 3¼-foot Risdon with its small daily capacity "a gopher to the Andes compared with the task and the material at hand," then went on to laud the capabilities of the huge new dredge.[59]

Both dredges worked with remarkable success from 1911 well into 1916, when they ran out of ground. They dug round the clock, with three shifts, which in the case of the Yuba involved four men per shift. The Risdon set a precedent in the basin that first year by digging continually throughout the winter season of 1910–11. Working on Elk Creek, the big Yuba did likewise. Cold weather severely strained electric cables, with snow and ice increasing

FIGURE 12.3. The flagship of the Boston and Idaho Gold Dredging Company. Courtesy, Idaho State Historical Society, Boise, ISHS 72-112.b.

the weight of the usual galvanized wire armor, making it more difficult to handle in the winter. Robert Bell extolled the virtues of that "marvel of mechanical, operation and efficiency," which in its first year had dug its way a full mile to Mores Creek, "completely eating up and turning over the deep gravel bed from rim to rim." In October the Yuba set a record of a different type when it processed a total of 361,876 yards. In 1912 it was averaging 307,623 yards per month while running 19.9 percent in downtime, according to superintendent John Miles. That gave the "reliable sources" crowd an opportunity to calculate and speculate. If the large machine was handling, say, an average of 8,000 cubic yards a day of gravel said to contain not less than 17 cents in gold per yard and that sometimes ran as high as 50 cents a yard, would not the monthly gold recovery be $40,000 or "practically $500,000 a year"?[60]

Placer gold output of Boise County was $528,257 in 1913, and the Boston and Idaho Gold Dredging Company "was by far the largest producer," having worked over 4 million yards during the year. State Inspector of Mines

FIGURE 12.4. The size of the Boston and Idaho Company's fifteen-feet bucket dredge is reflected by that of the worker standing beside the lower tumbler. Courtesy, Idaho State Historical Society, Boise, ISHS 72-112.b.

FIGURE 12.5. Not only was the Boston and Idaho Gold Dredging Company's metal monster successful, even when idle it was awesome to both adults and children. Courtesy, Idaho State Historical Society, Boise, ISHS 72-112.b.

Bell believed Boise Basin was probably responsible for nearly half of Idaho's gold output in 1913. The following year, with the 15-foot buckets replaced with those of 18-foot capacity, Estabrook's crew treated nearly 5 million yards of gravel and again led the county in gold output. In 1915 a cold winter limited the company's work to only ten months—and on poorer ground—and the big boat handled less than 3.5 million yards of material. According to Etheredge Walker, the company's secretary, its operating costs were 2.772 cents per yard. Although it dug less gold than it had the previous season, the company still led all others in the district.[61] Indeed, sometime during the year, with 18-foot buckets, it broke its old record for cubic yards dug in a single month. *Mining and Engineering World* quoted the figure of 400,000 yards. Nine years later Samuel Ballard put it even higher—461,000 yards.[62]

Perhaps the most sensational news concerning the company in 1916 came in large headlines on August 22: "BASIN BANDIT GETS CLEAN-UP OF BIG DREDGE." A day or two earlier, Estabrook had been performing one of his regular tasks, driving his automobile to pick up the gold amalgam from the dredge, which was working two miles from Idaho City. Normally, after each periodic cleanup he would take the gold to his office there and lock it in his safe to await collection of a larger amount to bring to the US Assay Office in Boise. On this occasion, when Estabrook stopped his car to remove brush and tree limbs from the lonely mountain road, he was accosted by a gunman, masked with a bandana, who proceeded to strip him not only of the gold but also of the $100 dollars in his wallet before departing. The local sheriff called out a posse, and the warden of the penitentiary brought his bloodhounds and traced the culprit six miles before losing the trail. No value of the gold appeared in the *Statesman*, but a newspaper in Ogden, Utah, wrote that it was worth $2,180.[63]

In mid-1916 Robert Bell reiterated that Estabrook's machine "continues to plow up the bed of More creek [*sic*]," although the remaining ground was of a low grade. Wrongly as it turned out, Bell believed Boston and Idaho's efficient management promised an extended corporate life. The county's gold output was down $158,845 from the amount the previous year, but the firm's big boat continued to top the production list.[64]

On the last day of 1916, the Boston and Idaho Gold Dredging Company officially announced that it had exhausted its ground. It reported that its worn bucket line would be renewed and that the boat would run on a small tract that was available. Because of the wartime shortage of material for repairs, the dredge remained idle until the very end of 1917. The rest of its working life was brief. For several years the company searched in vain for a new placer site, including drill prospecting on Crow Creek near Radersburg, Montana, and elaborate testing on upper Grimes Creek in the basin. By the end of 1918 the big boat was being dismantled for shipment to the firm's California property at Trinity Center.[65] Captain Winthrop Estabrook went off to France in 1917 with the Twentieth Engineers, but when the war was over he followed his dredge south to a more salubrious climate.[66]

The Boston and Idaho Gold Dredging Company retained some of its investments in the Boise Basin. It had acquired operating control of the Gold Hill lode deposits in Quartzburg—the oldest successful lode mine in the state, one that had been operated off and on for more than fifty years. The company also put considerable capital into the venerable Banner silver mine being successfully revived and further developed by Vivian Thorne. In addition, it kept its power plant in the South Fork of the Payette River, which provided electrical current to a number of towns and mines in the area.[67]

In the spring of 1922 an important mining deal was concluded in Pocatello. The Gold Dredging and Power Corporation, a Boise firm, took an option to buy basin property owned by Winthrop Estabrook and valued at $750,000. Included were over 5,800 acres of "rich placer ground," all remaining equipment of the Boston and Idaho Gold Dredging Company—which meant the hydroelectric plant—and the Centerville Mining and Milling Company's holdings on Grimes Creek. The Gold Dredging and Power Corporation had been capitalized at $1.75 million and was sponsored by Sheridan K. Atkinson, Frank E. Johnesse, W. D. Bohm, S.W. Thirkill, and Ernest G. Eagleson.[68]

Unfortunately, about a month later the power plant on the Payette River was completely destroyed by a fire caused by a workman trying to make a hole in the roof with a hot iron. Damages were estimated at $100,000 but were covered by insurance, and reconstruction began almost immediately. Whether the Gold Dredging and Power Corporation's option was affected was not stated. When the fire occurred, an expert was on his way from London to assess the property for an underwriting company prepared to back the buyer in the transaction.[69] Eventually, after several years of delay, Gold Dredging and Power put a dredge on Grimes Creek near Centerville.[70]

Thus ended the saga of the Boston and Idaho Gold Dredging Company. A fitting tribute came from Inspector of Mines Robert Bell at the time of the closure announcement: "This enterprise has proven the largest individual source of gold production for several years past. It has been one of the most substantially financed and cleverly operated mechanical mining enterprises in the State. Its equipment embraces one of the largest types of chain elevator bucket dredges in the country, which, while only involving a lift of about forty feet, has actually mined and concentrated as much metal bearing material in a year as all the lode mines of the State combined."[71]

Paralleling part of the time span of the Bostonians and successful in a small way was the Moline Mining Company. Unlike the Basic Company and its satellite concerns, the Moline Mining Company worked from 1902 through the 1911 season on Fall Creek and Granite Creek, where the firm owned eighty acres of ground about two miles southwest of Placerville. Total production was slightly more than $444,000, as its dredge chewed its way along about a mile of the Fall Creek channel. Incorporated in Illinois, the company was probably owned jointly by only three men, of whom H. A. Ainsworth of the town of Moline was the capitalist who provided much of the wherewithal.

Of the other two, E. F. Blaine was a veteran mining man generally regarded as the manager, although Robert Nye, an engineer who graduated from the Colorado School of Mines, was listed as holding that position in the early years of the operation. The *Engineering and Mining Journal* in 1904 named all three men as owners.[72]

At first, the company used a steam-powered dipper dredge mounted on tracks to lift the gravel into a hopper that fed into a set of rotary screens and then into sluice boxes with Hungarian riffles. The dredge could handle 1,100 cubic yards a day, with production averaging twenty-eight cents per yard; it made money with a crew of eighteen men. In 1904 the company's gold yield was higher than that of any other dredge in the state. In 1905 it abandoned the dipper, and Ainsworth purchased a 5-foot open-connected Risdon, dubbed *Granite No. 1*. In March of that year a carload of Oregon fir timbers had arrived in Boise, the first of eighteen railroad cars on the way from Risdon Iron Works. Risdon also sent mechanics from California to assemble the big machine, which was expected to be operational by fall. Steam-powered, the new boat could reach a depth of 30 feet and by working three shifts could handle 1,400 yards a day—somewhat less than the 2,000 yards earlier predicted.[73]

The "Blaine dredge," a term often used, seems not to have done much digging in 1905. In late August it was reported to be "running smoothly," and the records show it had been active on Corcoran bar. In his *Annual Report*, Robert Bell remarked that the Moline Mining Company's dredge had operated successfully for several months of the next season, "but the work was seriously interrupted by accidents to important parts of the machinery." "Hard luck," Bell called it. When it started up on April 4, the dredge worked well until May 23, when the engine shaft broke. It could not be repaired in the basin; the engine was of the compound type, with two cranks on the shaft, and the latter had to be forged new at Risdon Iron Works. Risdon did not complete the work and ship the new part for two weeks, much to the consternation of Blaine, who knew the negative cost of downtime. Apparently, this was not the only interruption, but Bell did not elaborate further. The *Granite No. 1* used cocoa mats to catch the gold. Woven very fine, they held the gold but took too long to clean. They had to be removed and washed in tubs of water. Blaine contemplated amalgamation after installing metal riffles and using mercury to catch the precious metal, as in ordinary placer mining. The ground worked by the boat ranged from seventeen feet to twenty-two feet in depth and was overlaid by tailings from earlier mining, which had to be removed but which also meant a small bonus from them, averaging three cents a yard.[74]

From 1905 to 1908 the *Granite No. 1* was the only bucket-line dredge working in the basin. After its mechanical problems in 1906, it began an unspectacular, steady, and profitable run. Several years in a row, it was the largest producer of placer gold in the basin. A laudatory article in *Mining and Scientific Press* at the beginning of 1910 noted that for the past four-and-a half years the dredge had dug on average eight months a season, twenty-seven days a month. The 1909 season had been shorter than usual because the boat had been dry-docked for an overhaul and reinforcement of the hull—an unusual kind of repair.[75]

In August 1910 Blaine commented on the "very satisfactory" run the dredge was having. He pointed out that the Moline Mining Company still owned a substantial area of virgin ground and that the length of time required to work it would be longer than the amount of time the present dredge would last. A Utah newspaper carried the story under the headline "WILL NEED NEW DREDGE." Blaine was wrong in his assessment of the situation. *Granite No. 1* completed the 1911 season, then closed down for good. Increasing mechanical difficulties as the boat approached Fall Creek Canyon, combined with deeper, unreachable bedrock, spelled the end. E. T. Blaine immediately took over management of the Boise Basin Improvement Company's hydraulic mining operation near Idaho City.[76]

No bucket dredges operated in the Boise Basin between 1916 and 1926. In 1912 Stephen M. Smith incorporated the Mores Creek Boulder Gold Dredging Company. Described by the press as a "miner of the Boise Basin country" and as "Doctor" by some of his colleagues in the American Mining Congress, Smith was an inventor who believed he had found a way to mine an eight-mile virgin stretch of Mores Creek that had never been systematically worked because of the presence of large boulders. His company, nominally capitalized at $150,000, was formed to utilize a new electric aerial cable system, whereby fifteen men working in three shifts of five each would be able to handle about 4,000 cubic yards every twenty-four hours.[77]

Nothing happened at the time. Three years later, at a meeting of the American Mining Congress in San Francisco, Smith described one of his recent inventions—a grapple especially adapted for handling boulders of different sizes and shapes. Using such equipment, he claimed, a man with a derrick or a boom could pick up boulders and drop them wherever he pleased, without a sling or chains.[78] In 1916 a Chicago mining publication noted that Smith's Mores Creek Boulder Gold Dredging Company was building a five-and-a-half-foot bucket dredge.[79] There was still no evidence of actual mining and no further mention of the concern until 1920, when the *American Mining Manual* listed it as having a steam-powered gold dredge and employing fifteen men, which may have been mere window dressing. Two years later the annual report of the Idaho inspector of mines indicated that the company had issued 353,611 of its shares, had eight unpatented claims on Mores Creek, had a "partly completed dredge" and an aerial cable system, but employed only two men. The inspector added that the firm also owned the Highland Dam power plant, from whence power could be transmitted eight miles to the dredge. By 1923, the next year, the boat had been dismantled and the company dissolved.[80]

For several years, litigation kept the Consolidated Mines Syndicate from dredging disputed ground on Grimes Creek, but in 1925 legal matters were settled, and the company gained full possession of the property. The syndicate promptly tapped a subordinate firm, the Gold Dredging and Power Corporation, to erect and operate a four-foot steam-driven Risdon on Grimes Creek near Centerville. Frank E. Johnesse, a self-made mining engineer, was president of the syndicate, while Sheridan K. Atkinson, a graduate of the University of Wisconsin, headed the Gold Dredging and Power Corporation and was manager in the field.[81]

The Consolidated Mines Syndicate owned four miles on Grimes Creek. The boat commenced digging on June 1, 1926, under the supervision of George Carr, former superintendent of the Yuba Consolidated Gold Fields in California, where he had overseen the work of twelve dredges. By mid-December, when the dredge closed down for the winter, it had handled 483,000 cubic yards of gravel, an auspicious beginning. In early 1927 it was converted to electric power. Sometime the next year, the Gold Dredging and Power Corporation was reorganized and became the Idaho Gold Dredging Corporation.[82]

Under whatever name, during its half-dozen years of operation, the boat was a consistent producer. In its partial run in 1926, it was one of only four enterprises in the state with an output of more than $20,000 in gold. Three of those were dredging firms. In 1927 it was one of three Idaho operations—two of which were dredge miners—to produce more than $30,000. The next season saw the Idaho Gold Dredging Corporation's take drop considerably, although it was still a respectable $24,000 plus. In 1929 the machine worked 262 days and recovered gold valued at about $37,000. The next season was even better—in 257 days of digging, the bullion retrieved was worth $49,000.[83]

In early 1928 the company acquired new placer ground on the watershed of Grimes Creek previously owned by the Boise Payette Lumber Company. From 1920 to 1926, Boise Payette had cut the timber. In so doing, the firm used about thirty-five miles of log chutes along the hillsides and gulches. To facilitate skidding of the logs, the lumber company had lubricated about half of the chutes with grease and oil—much of which dropped off, saturated the ground, soaked into the trash and bark that accumulated in and near Grimes Creek's tributaries, and found its way into the dredge pond. As a result, the gold and mercury were so contaminated that they would not amalgamate, and most of the precious metal was lost. Atkinson estimated that on the basis of pre-dredge drilling, the company had a loss of precisely $389,369.45. Idaho Gold Dredging Corporation brought suit and ultimately won a $100,000 judgment from Boise Payette.[84]

The year 1931 brought disaster. On July 13 the company's dredge was completely destroyed by fire, probably caused by a short circuit in the electrical system. The boat had been in operation for six years and was so thoroughly dried out that nothing could be saved. Not only was there a large loss in machinery, but being forced to halt digging in mid-season also meant a substantial loss in income to the owners of about $36,000 compared with the previous year. The disaster was partially covered by insurance. Atkinson and his directors quickly made plans for a replacement dredge, but the company produced no gold in 1932.[85]

Several new dredging efforts coincided with part of the Gold Dredging and Power Corporation–Idaho Gold Dredging Corporation run. In the spring of 1927, it was announced that a new dredge would be started up by Melvin T. Rowland by about May 1. Later Spokane newspapers were more specific. Rowland, a resident of Nampa with a mining interest in the Owyhee area, was having the requisite machinery delivered to the Kempner ranch on the Crooked River in Boise County, twenty-two miles northeast of Idaho City. This machine was described briefly as a three-foot bucket-line dredge of the "sluice box type," which meant in the language of the trade a "flume dredge," and it was a throwback to an earlier era. On this kind of boat, hopper, screen tables, and stacker were replaced by a long flume running down the center of the vessel far enough astern to dump the tailings away from the pool. The buckets dumped directly into the flume, and as water washed over its riffles, the gold was caught with quicksilver. The flume dredge was smaller than the standard dredge, relatively cheap, and especially adapted to work narrow, rich pay streaks with less water. It was also a good deal more inefficient and lost more of the values than the standard dredge. The machine of Rowland and associates was powered by two sixty horsepower Fairbanks-Morse diesel engines and had been built by the Straub Manufacturing Company of Oakland, California, a firm well-known for turning out small dredges for Alaska prior to World War I. Estimates of the price of the boat varied from $65,000 to $92,000 in stories that reached the public, but whatever the cost, it was underwritten by the McLaren brothers, Walter and Alexander, wealthy inventors of an ice cream cone machine. The Crooked River Mining Company was incorporated in early October the same year, with offices in Dayton, Ohio, also the home of the McLaren Consolidated Cone Corporation. Rowland was the company's president and manager, but he issued no shares, and the firm subsequently forfeited its charter in late 1931. In December 1927 the *Salt Lake Mining Review* recorded a visit from Rowland and observed that he and his associates were equipping "a powerful dredge" for the Eldorado Gravel Company to be used in the upper Boise Basin placers the following spring.[86]

The machine was a catastrophe and sank after it was launched. When it was raised, sluice boxes were substituted for the flume at a cost of $60,000 in 1928, but there was no production that year. During the next season the boat ran for twenty-seven days and handled 30,000 cubic yards of gravel. In 1930 it dug for forty-six days, treated twice that amount of gravel, and had an output of $5,500 in gold. Rowland ran the Crooked River dredge during part of 1931 and for a short time the following year.[87] That was the end of the unfortunate boat's working life.

In 1933 a rich piece of placer ground became available in the basin—the sizable acreage on Grimes, Granite, and Wolf Creeks that had gone to Joseph Wharton in 1904 as part of the settlement of the two mortgages he held with the Basic Company. At the time, Wharton did nothing with the property. In addition, his estate took no action for many years after his death, although it was well-known that the ground was available—at a high price. In late December 1912 Oscar C. Perry, Guggenheim's chief engineer, wrote a colleague about approaching the executor of the estate, Harrison S. Morris. Would it be better to go to Philadelphia and try to make a deal directly with the executor, or would it be more effective to negotiate through some other channel? "Would the magic name of Guggenheim cause these people to stand by the original price of $600,000.00 for the property, or do you think they would at this time deal on a basis of $300,000.00 or less," Perry asked.[88]

Whether any approach was made at this time is uncertain, but the ground remained dormant until 1933, when the Wharton estate optioned it to Edward E. Marshall, longtime president of the American Manganese Manufacturing Company of Philadelphia. Marshall, in turn, leased seven-and-a-half miles on Grimes, Granite, and Wolf Creeks to a pair of Texas contractors, H. F. McFarland and Louis Whithan, who brought in three draglines and a washing plant. With this equipment, the two worked near Placerville through the summer of 1933 and well into 1934, when the outfit ranked fifth in the state in gold production.[89]

When the US Congress raised the price of gold to thirty-five dollars an ounce in 1935, it signaled the beginning of a new era for Boise Basin. As one mining man wrote to another in 1937, "You would be interested to know that in the immediate vicinity of Idaho City three new dredges have been constructed and very profitably operated this past season, together with some drag line, sluice and hydraulic placer operations." In fact, the basin showed much real activity in both placer and lode developments. Nine bucket dredges were working in Idaho in 1935, compared with five in 1934; in 1937 there were ten.[90] This boom continued from 1934 to October 15, 1942, when the War Production Board's order to halt all gold mining until further notice took effect. Some of the work was performed by newcomers; others were seasoned veterans with years of dredging behind them.

One of the most successful moves into the basin was made by Fred Baumhoff and Al Fisher. These co-partners had done extremely well over the past few years at Warren. In December 1934 they were undoubtedly thinking of relocating when they requested cost figures from the Yuba Manufacturing Company for a six-foot electric steel-hulled dredge with a bucket line for digging thirty feet. Yuba responded that it would furnish all of the equipment for such a machine; the partners would furnish their own hull, trusses, gantries, and stacker system. The manufacturer would deliver FOB to any designated point in Idaho for $92,546, plus the actual cost of freight from Benitia, California. The price would be split into seven payments before January 4, 1935, with a delivery date of May 15 the same year.[91]

This plan was not acted upon at the time, but in 1935 Baumhoff and Fisher did purchase a six-foot Yuba (No. 106) for use in Boise Basin. Since it came with a steel hull, it must have cost a good deal more than the 1934 estimate. Launched in mid-July, the machine was the largest dredge in the state at that time and certainly one of the most profitable. Even in its truncated maiden season it ranked fifth in the state in gold production.[92]

The Grimes Company was also new to the basin, although its president, Henry L. Day, the owner of the Hercules Mine in the Coeur d'Alene district, had a wealth of hard-rock mining experience behind him. The Murphy brothers, George and Harry, were also partners. The Grimes Creek dredge had a history; it had originally been built for the Bed Rock Dredging Company and had then been owned by the Idaho Gold Dredging Corporation until it was destroyed by fire in 1931. Now, in 1935, it was rebuilt on Grimes Creek near Pioneerville and put to work under the supervision of dredgemaster Miles Young. Immediately, it became the second-largest producer in the basin.[93]

Another boat started up in 1936, operated by the Mores Creek Dredging Company, a firm organized by George E. Murphy and R. H. Ellis. The firm controlled 640 acres of patented and 140 acres of unpatented property running along Mores Creek from about three miles below Warm Springs to about two miles above Idaho City, plus a 40-acre tract on Elk Creek—all held under lease and option. Below Warm Springs, where the boat began working, the ground had previously been worked by hand methods only. Much of the gold was in the form of amalgam lost by the early miners and was distributed fairly uniformly from surface to bedrock. The dredge itself was of timber construction and electrically driven, with five-foot buckets when was brought in from the Salmon area. Later, the buckets were redesigned by dredgemaster J. D. Smith of Idaho City, who extended the lips seven inches beyond the ends of the pins to ease wear on the lower tumbler when crowding into the bank to make a cut. After the change, the buckets were of eight-foot capacity. In early September 1937 the *Statesman* ran a photograph of an employee of First Security Bank holding a gold ingot valued at $14,000, with a caption telling readers that two similar bars were brought into the bank each month by the Mores Creek Dredging Company.[94]

Following the destruction of its dredge by fire in 1931, the Idaho Gold Dredging Corporation seems to have dropped out of sight for a few years. In 1937 it resurfaced, when the *Spokane Chronicle* published a picture of one of its two boats and a brief story about both. The one depicted was handling about 3,500 yards of material a day on Grimes Creek. The other was on Mores Creek, four miles below Idaho City, where it was treating 5,000 yards a day in gravel that averaged twenty-three-and-a-half cents per yard. At the same time, Sheridan K. Atkinson announced that the company had acquired new placer property near Atlanta and was also developing a gold-quartz mine near Murphy, in Owyhee County.[95] It is strange that nothing appeared in the US Geological Survey's *Minerals Yearbooks* between 1932 and 1940 to indicate that the company was actively engaged in dredging again after 1931.

Draglines appeared and disappeared. One of the first was the Granite Creek Dredging Company, organized by W. L. Smith, president, M. M. Smith, and John Whitney, all from Boise. This group put a dragline and washing plant on Fall Creek near Placerville in the summer of 1935. The press reported that a bucket-line machine—the Blaine dredge—had previously worked up the stream as far as it was possible to float and recovered twenty-nine cents a cubic yard at the old price, which meant about fifty cents a yard in 1935. Since the placer deposit was large, it was believed that operations "may continue for years." That proved to be a poor prediction. The machine ran briefly, but the ground was leased to Fisher and Baumhoff in 1936, and the Granite Creek Dredging Company moved its equipment to Florence.[96]

During the summer and fall of 1937, Harry England, a veteran dragline operator from Oroville, California, built two of his "doodlebugs," as they were often called—one on Grimes Creek above Pioneerville and one on a tributary of Fall Creek, west of the town of Granite. On Grimes Creek, where a smaller one-yard diesel-powered machine was working the deposits mined earlier by hand, there was relatively little overburden, but there was heavy growth of brush and evergreens, which required a bulldozer to clear. Here, the washing plant was set on steel pontoons and driven from a line shaft by a separate diesel engine. On the Fall Creek feeder, the digging was done by a three-yard dragline with a sixty-five-foot boom in gravel about fifteen feet deep and covered with five feet of hard topsoil, which proved difficult to excavate. Power for the large rig was provided by a diesel electric generating plant installed on the boat. Between the two dredges, England and his associates—O. M. Warren, W. A. Hilliard, and Lord and Bishop of Sacramento—were employing thirty-two men. The labor necessary to run the large machine three shifts a day included one dredgemaster, three dragline operators, three washing-plant men, three oilers, and two shoremen. The shoremen cleared the ground and did other miscellaneous tasks. The versatile Lord and Bishop was actually a Sacramento contracting firm specializing in roads and bridges, but in this instance the firm ran the large three-yard operation in the Fall Creek area through 1938, when it moved its machinery to California. At the same time, Harry England relocated his equipment to Baker, Oregon.[97]

By 1941, two more dragline floating dredges were in Boise Basin, one operated by M. A. Stickler on Mores Creek and the other by the Smith Brothers on Grimes Creek.[98] In 1942, when all gold mining was prohibited, both closed down and did not reappear when the industry resumed production.

It was not the draglines that were so important to the basin's economy in the period 1934 to 1942, it was the four bucket dredges—two run by Fisher and Baumhoff and one each by the Grimes Company and the Mores Creek Dredging Company–Idaho-Canadian Dredging Company. Perhaps it was six boats, if the two phantom dredges of the Idaho Gold Dredging Corporation are counted. Bucket dredges were able to produce much more than the draglines. For example, in 1937, bucket-line boats in the basin had an output of 17,438 fine ounces of gold and 4,386 ounces of silver, whereas dragline operations recovered only 1,193 fine ounces of gold and 355 ounces of silver. For the following year the figures were 18,549 fine ounces of gold and 4,553 ounces of silver taken out by bucket dredges and 2,139 ounces of gold and 529 ounces of silver recovered by dragline.[99]

In 1935 Fisher and Baumhoff"s six-foot Yuba set the pace on Grimes Creek, working ground on the Wharton Estate near Centerville. During the next year they added the two-and-a-half-foot *Mickey Mouse,* trucking it from Warren and rebuilding it on Granite Creek. The small boat was digging by the end of July on a round-the-clock basis, with eight men, while the big Yuba had a comparable-size crew plus a mechanic and two extras. Both boats were powered by the Grimes Pass Power Company, which owned the plant on the South Fork of the Payette River built in 1910 by the Boston and Idaho Gold

FIGURE 12.6. Fred Baumhoff clears the ground cover from gravel yet to be dredged in Boise Basin. Courtesy, Idaho State Historical Society, Boise, ISHS 72-112.b.

Dredging Company. The *Mickey Mouse* worked only four months in 1936, dug 170,000 cubic yards, and recovered gold valued at $20,08.66 and silver worth $78.76, for a loss of $3,224.24. In contrast, the big boat at Centerville handled 1,596,000 cubic yards of gravel that produced gold appraised at $213,066 and silver valued at $1,134.29, for a net profit of $99,500.04 figured on the basis of direct operating costs. General and overhead costs amounting to $8,640.28 were not included. When these costs were subtracted, the total profit for the season was $90,859.76.[100]

In 1937 the Boise Basin district was the leading producer of gold in Idaho, with a total of 25,653 fine ounces. Of that amount, 19,724 ounces came from placer mining. The four bucket dredges in the basin contributed the bulk of that amount—17,438 ounces. The two Fisher-Baumhoff boats had the largest output in the state that year, while the Mores Creek Dredging Company came in third. The Grimes Company "was also a large producer," and its boat was busy in May of that year ravaging the old mining town of Pioneerville. A number of buildings had to be moved to get at the rich ground underneath them, and the dredge encountered some of the old Chinese dens as the digging progressed. The Grimes Company boat worked through 1939, when it ran out of ground; it was then moved to Oregon.[101]

The Fisher-Baumhoff dredges continued to be the largest gold producers in Idaho in 1937, 1938, and 1939, but they fell to second place in 1940 and 1941 behind Talache Mines, Inc., a lode mining operation in Atlanta, although they remained the top producer of placer gold in the state. In 1940 the company's two boats continued to dig on Granite and Grimes Creeks and during the year dug at the town site of Old Centerville to Grimes Creek, which would be worked during the spring by the big Yuba.[102]

When Yuba's follow-up engineer, Edward A. Foster, visited the new boat just purchased by Boise King Placers on the Middle Fork of the Boise River in the spring of 1941, he also looked in on the Baumhoff and Fisher dredge at Centerville. He found the latter in excellent condition, with the standard gold-saving tables having been rebuilt in the field to also use jigs—a modern and more efficient innovation just coming into use. Foster noted that during the past year the dredge had made a gain of about 400 pounds of quicksilver as a result of digging in old placer developments.[103]

The Mores Creek Dredging Company was also active and productive before selling its buildings, grounds, and dredge to the Idaho-Canadian Dredging Company in early May 1939 for a pittance. There was a striking similarity among the executive personnel of Mores Creek Dredging, the Grimes Company, and the new firm, of which Harry Murphy was president and brother George was secretary-treasurer, while Miles Young served as dredgemaster as well as a corporate director.[104]

Obviously, all four dredges had prospered. By mid-October 1942, when the War Production Board closed down all nonessential mining, Fred Baumhoff and Al Fisher were wealthy men. In 1941 Fisher commissioned the well-known architects Tourtellete and Hummel to build him a home on Boise's prestigious Harrison Boulevard. Five years later he sold the house for $35,000 to Jack Simplot, who had become wealthy from both potatoes and mines. In 2011 the house was again up for sale, at the asking price of $1.5 million.[105]

California native Fred Baumhoff applied part of his Idaho-earned fortune to further his ambition of trying to turn the classic old Strawberry Lodge, seventeen miles west of Lake Tahoe, into a posh ski resort—one that, in his words, would "wipe Sun Valley off the map." Strawberry Resort opened for business over Memorial weekend 1940. But the coming of World War II, with its gas, tire, and steel rationing, abruptly halted Baumhoff's grandiose plans. He sold his California investment in 1942 and returned to Idaho, where he resumed mining and won a seat in the state senate. Guests today at the historic lodge just south of Echo Summit still walk by a decrepit chairlift tower near the parking lot, a quiet reminder of Baumhoff's shattered dream.[106]

Fisher and Baumhoff were ambitious and eager to expand their operations. To do so, as early as 1936, when they moved to Boise Basin, they sometimes played a kind of musical dredges game, moving equipment to other mining districts around the state. It was well-known that the two men had tested gravels along the American River in the Elk City district and bought up all the wild hay meadows in Elk City itself.[107] In 1938 they installed a four-foot bucket-line dredge on Little Smoky Creek but discontinued it in 1939.[108] A dredge labeled the "Baumhoff, Fisher and McDowell" was pulled out of Warren ground in August 1939 and moved to Moose Creek near Salmon City. There, it began operation on the McNutt property in June the following year under the auspices of Fisher and Higgins. How long it worked was not recorded.[109]

In the last quarter of the following year, a four-foot bucket dredge was put into operation by the Idaho-Warren Dredging Company at the Hughes Creek Placer in the Gibbonsville district, in far eastern Lemhi County. The machine recovered 1,900 ounces of gold and 203 ounces of silver in 1942, before being closed by the wartime prohibition on gold mining. It did not get back into action until 1947, and then it worked for less than two years.[110]

When the wartime ban went into effect, both Fisher and Baumhoff had enough ground left in Boise Basin for several more years of runs, but it was some distance away. Wartime priorities made machinery and repairs impossible, so the partners took no action to work or move the machine.[111] The ban had been rescinded on July 1, 1945, two months before the Japanese surrendered. In Boise Basin the first to resume dredging was the Idaho-Canadian Dredging Company, which started up its seven-and-a-half-foot boat again in September near Idaho City. In March 1946 Fred Baumhoff was back on Grimes Creek with the six-foot Yuba as well as a new partner. Now the company was called Baumhoff-Marshall, Inc. The *Mickey Mouse* started in March on Granite Creek, but before the end of the season it was moved to Wolf Creek and rebuilt. The company recovered 4,943 fine ounces of gold and 1,111 ounces of silver that year. Idaho-Canadian closed down repairs in

FIGURE 12.7. Fred Baumhoff in legislative garb. With Al Fisher, he made a powerful team, with dredges operating at various times in different parts of the state. Courtesy, Idaho State Historical Society, Boise, ISHS 72-112.b.

January 1946 and did not resume digging until March. Digging continued until October 9 when the Yuba capsized, yet the company still accounted for 2,602 ounces of gold and 627 ounces of silver that season. At the same time, a newcomer appeared—Pyramid Mining Company, with a bucket dredge near Twin Springs.[112]

The Pyramid Mining Company never reappeared the next season, but the Idaho-Canadian Dredging Company was back in 1947 with a different rig. It had abandoned its old seven-and-a-half-foot dredge in favor of a newer six-foot machine it had procured from Bearmouth, Montana. Baumhoff-Marshall, Inc.'s, two boats operated on Granite Creek near Centerville and treated about a million cubic yards of gravel. In 1948 Baumhoff-Marshall apparently ran only the big Yuba for the first seven months of the year, while Idaho-Canadian dug the entire season on Mores Creek. Combined, the two boats produced a total of 11,732 ounces of gold and 4,466 ounces of silver.[113]

In March 1949 Baumhoff-Marshall, Inc., suspended work because of poor ground. For fifteen years the Baumhoff-Fisher or Baumhoff-Marshall combination had successfully dug gold in the basin. The larger dredge was dismantled and sent to Cascade, where it became part of the federal drive to build a stockpile of rare minerals.[114] Another boat belonging to the same owners was also shifted from Warren for the same purpose. What happened to the *Mickey Mouse* is uncertain.

Idaho-Canadian continued to operate on Mores Creek, and its dredge was the only one active in the basin; thus, it constituted the main source of gold produced therein. Digging from March 15 to October 7, the boat treated 1.15 million cubic yards of gravel in 1949; with a late March start in 1950, it handled 960,000 yards of material. In its final, truncated season, from January to July 31, 1951, the six-foot Yuba washed 702,000 yards. In its last three years, the machine recovered a total of 21,346 fine ounces of gold and 11,263 ounces of silver. In 1952 it was disassembled and moved to Cascade to dig rare minerals.[115]

Boise Basin was undoubtedly the richest placer mining area in Idaho, producing an estimated 98 percent of the gold mined in Boise County. How much the dredges accounted for is impossible to say, but from the beginning of the twentieth century they were responsible for a sizable portion. There was a time, not very long ago, when visitors entering the town of Idaho City were greeted with a large sign proclaiming that "Boise Basin Produced More Gold than All of Alaska," a catch phrase echoed by a few popular writers.[116] Facts and figures do not bear out such a statement, however. According to the best estimates, mining in Boise Basin had a total output of about 2.9 million troy ounces of gold in the period 1862 to 1969. In comparison, government geologists put Alaska production at nearly 29 million troy ounces between 1880 and 1957, and other experts cite a figure of 40.3 million troy ounces between 1880 and the end of 2007.[117] Idaho's total bucket dredge recovery from 1896 through 1956 came to 719,158 ounces, placing the state fourth in the United States in that category, after California, Alaska, and Montana.[118]

Notes

1. James L. Onderdonk, *Idaho Facts and Statistics Concerning Its Mining, Farming, Stock-Raising, Lumbering, and Other Resources and Industries* (San Francisco: A. L. Bancroft, 1885), 37; William S. Greever, *Bonanza West: The Story of the Western Mining Rushes, 1848–1900* (Boise: University of Idaho Press, 1963), 262–63; Merle W. Wells, *Gold Camps and Silver Cities*, Bulletin 22, 2nd ed. (Moscow: Idaho Department of Lands, Bureau of Mines and Geology, 1983), 2–3, 5.

2. Hubert Howe Bancroft, *The Works of Hubert Howe Bancroft*, vol. 31 (San Francisco: History Company, 1890), 261, 427, 434; Greever, *Bonanza West*, 265.

3. William J. Trimble, *The Mining Advance into the Inland Empire* (Madison: University of Wisconsin Press, 1914), 75–76.

4. Ibid.

5. Waldemar Lindgren, *The Mining Districts of the Idaho Basin and the Boise Ridge* (Washington, DC: Government Printing Office, 1898), 651; Bancroft, *The Works*, 31:261; Samuel M. Ballard, "Geology and Gold Resources of Boise Basin, Boise County Idaho," Idaho Bureau of Mines and Geology Bulletin no. 9 (Moscow: University of Idaho, December 1924), 14; Walter J. Bawiec and Lawrence L. Dee, "Historical Perspective on the Mineral Production of Idaho, with Comments on the Hailey 1°x2° Quadrangle, Idaho," US Geological Survey Bulletin 2064-T, Version 1.0 (Denver: US Geological Survey, 2001), 18; George Crofutt, *Great Trans-Continental Tourist's Guide* (New York: Geo. A. Crofutt, 1870), 117.

6. Elijah Chaney, "Boise Basin," *Ballou's Monthly Magazine* (Boston) 37 (January 1873): 52.

7. A. H. Koschmann and M. H. Bergendahl, "Principal Gold-Producing Districts of the United States," Geological Survey Professional Paper 610 (Washington, DC: Government Printing Office, 1968), 123–25; Thor H. Kiilsgaard, Frederick S. Fisher, and Earl Bennett, "Gold-Silver Deposits Associated with the Trans-Challis Fault System, Idaho," US Geological Survey Bulletin 1857 (n.d.), chapter B, part 1, b25.

8. Chaney, "Boise Basin," 52; Ballard, "Geology and Gold Resources of Boise Basin," 17.

9. "Census of 1863," Idaho State Historical Society Reference Series 129, n.p., Idaho State Historical Society, Boise; "Census of 1964," Idaho State Historical Society Reference Series 130, n.p., Idaho State Historical Society, Boise; Wayne C. Sperling, *Southern Idaho Ghost Towns* (Caldwell, ID: Caxton, 1974), 45.

10. Lindgren, *Mining Districts of the Idaho Basin and the Boise Ridge*, 652, 655–56.

11. Ibid., 658; *Idaho Daily Statesman* (Boise), November 27, December 4, 2011.

12. Sharon Murray, "A Bucket Full Boise Basin's Dredging Heritage, 1898–1951," *Idaho Yesterdays* 34 (Fall 1990): 2; *Tenth Annual Catalogue of Swarthmore College, Swarthmore, Pa., 1878–79* (Philadelphia: Press of Friends Book Association, 1879), 41; *The 1919 Bucks County Intelligencer Obituary Index, Bucks County Intelligencer* Archives, n.p.; *Thirty-First Annual Catalogue of Swarthmore College Swarthmore Pa. 1899–1900* (Philadelphia: Franklin Printing, 1900), 4–5; for Wharton, see W. Ross Yates, *Joseph Wharton: Quaker Industrial Pioneer* (Bethlehem, PA; Lehigh University Press, 1987).

13. "Abstracts of Corporations," in *Acts of the Legislature of West Virginia at Its Twenty-Fourth Regular Session, Commencing January 11th, 1899* (Charleston, WV: Press Butler, 1899), 26, 30.

14. "Abstracts of Certificates of Incorporation Issued by the Secretary of State, and Not Before Reported," in ibid., 36, 191; John W. Leonard, ed., *Who's Who in America 1899–1900*, 1:37 (Chicago: A. N. Marquis, 1899).

15. *Idaho Daily Statesman* (Boise), January 20, April 7, 1898.

16. Ibid., April 14 and 15, 1898.

17. Ibid., June 2, 1899; *Engineering and Mining Journal* 64 (December 4, 1897): 663.

18. *Engineering and Mining Journal* 65 (June 18, 1898): 741; *Idaho Daily Statesman* (Boise), October 28, 1898.

19. *Engineering and Mining Journal* 66 (August 6, 1898): 166; *Engineering and Mining Journal* 68 (December 2, 1899): 669; *Mining and Scientific Press* 77 (July 30, 1898): 110; *Mining and Scientific Press* 80 (March 24, 1900): 317; *Idaho Daily Statesman* (Boise), September 1, October 28, 1898; *Salt Lake Herald*, July 25, August 17, 1898; Bucyrus Company, "List of Dredges," 1000075A, author's personal collection; Murray, "Bucket Full," 3.

20. "Report on the Grimes Creek Placers, Centerville, Idaho," December 21, 1912, typed copy, Fred L. Morris MSS, Bancroft Library, University of California, Berkeley (hereafter Morris MSS).

21. *Mining and Scientific Press* 79 (August 5, 1899): 149.

22. Ibid. 105 (September 14, 1912): 330; ibid. 81 (October 6, 1900): 400; ibid. 88 (March 1904): 25; Murray, "Bucket Full," 3–4.

23. *Idaho Daily Statesman* (Boise), January 22, November 9, 1899; January 1, March 1, 1902.

24. "Report on the Grimes Creek Placers, Centerville, Idaho," December 21, 1912, typed copy, Morris MSS; *Idaho Daily Statesman* (Boise), July 9, 1901; January 1, 1902.

25. *Idaho Daily Statesman* (Boise), March 1, 1902; *Engineering and Mining Journal* 78 (August 25, 1904): 297.

26. *Idaho Daily Statesman* (Boise), June 2, 1899.

27. Ibid., May 18, 1903; December 8, 1904; Hawley, *History of Idaho*, 3:257–58.

28. Robert N. Bell, *Eighth Annual Report of the Mining Industry of Idaho for the Year 1906* (State Mine Inspector's Report), 39; *Daily Statesman* (Boise), March 17, 1907; July 16, 1908.

29. Yates, *Joseph Wharton*, 340, 359.

30. Ibid., 358; Hawley, *History of Idaho*, 3:257–58.

31. *Idaho Daily Statesman* (Boise), December 9, 1907; "*Herman Hoopes et al. v. Basic Company*" (February 22, 1907), 72 *New Jersey Equity Reports*, 426.

32. *Salt Lake Tribune*, November 24, 1897; *Idaho Daily Statesman* (Boise), October 24, 1997.

33. C. Bancroft Gillespie, comp., *Illustrated History of South Boston* (South Boston: Inquirer, 1900), 110; *Iron Age* 88 (September 21, 1911): 625; *Railway Age* 39 (January 13, 1905): 59; *History of Boston Harbor 1891–1907*, 2 (n.p.: n.d.), 22, 223, 624, 847.

34. See "*San Diego Flume Company v. Souther et al.*," *Federal Reporter* 90 (October 3, 1898): 171.

35. *Engineering and Mining Journal* 112 (October 22, 1921): 646; *Idaho Daily Statesman* (Boise), January 20, 1898.

36. *The Hawes School Memorial* (Boston: David Clapp and Son, 1889), 4, 50; Murray, "Bucket Full," 3–4.

37. In 1899 the Boise *Statesman* indicated that the dredge cost $38,000; the next year the newspaper used the figure $40,000. *Idaho Daily Statesman*, September 16, 1899; June 27, 1900.

38. Murray, "Bucket Full," 3; George Hurst handwritten notebook, n.d., Box 4, Hurst MSS, Stanford University Archives, Palo Alto, CA (hereafter Hurst MSS); *Idaho Daily Statesman*, April 27 and 28, May 23 and 28, 1898; September 16, 1899.

39. *Idaho Daily Statesman* (Boise), August 2, September 1, 1898; December 6, 1902.

40. Ibid., November 19, 1898; March 31, April 16, 1899.

41. Ibid., April 26, 1899; *Bulletin of the American Institute of Mining Engineers* 100 (April 1915): xii.

42. *Idaho Daily Statesman* (Boise), September 16, 1899; June 27, 1900.

43. Ibid., October 8, 1903; Murray, "Bucket Full," 4; *Engineering and Mining Journal* 70 (September 15, 1900): 318; *Engineering and Mining Journal* 78 (August 25, 1904): 297.

44. *Record-Union* (Sacramento), April 30, 1899; *Idaho Daily Statesman* (Boise), September 16, 1899; March 14, 1901; October 29, 1905; September 13, 1909.

45. *Idaho Daily Statesman* (Boise), April 26, 1899; September 10, November 1, 1901.

46. Ibid., April 10, 1902.

47. Ibid., October 8, 1903.

48. Ibid.

49. *Railway Age* 39 (January 13, 1905): 59; *Iron Age* 88 (September 21, 1911): 625.

50. *Idaho Daily Statesman* (Boise), March 15, October 29, 1905.

51. Murray, "Bucket Full," 5; *Harvard Alumni Bulletin* 22 (December 18, 1919): 306.

52. *Idaho Daily Statesman* (Boise), March 3, 1912.

53. Ibid., June 26, 1906; Murray, "Bucket Full," 5.

54. Murray, "Bucket Full," 5; *Idaho Daily Statesman* (Boise), September 22, 1908; September 8, 1909; May 12, 1910; March 3, 1912; *Engineering and Mining Journal* 104 (January 13, 1912): 101; *Mining and Scientific Press* 105 (September 14, 1912): 330, 332;

F. Cushing Moore, *Eleventh Annual Report of the Mining Industry of Idaho for the Year 1909* (State Mine Inspector's Report), 86.

55. O. H. Metzger, "Reconnaissance of Placer Mining in Boise County, Idaho," US Bureau of Mines Information Circular 7028 (August 1938): 5.

56. Bucyrus Company, "List of Dredges," 100075C; Samuel Johnston, *"Gold Dredges": The Part Played by the Yuba Manufacturing Company in the Great Romance for Placer Gold, Platinum and Tin* (San Francisco: John Henry Nash, 1927), 41; *Idaho Daily Statesman* (Boise), August 25, 1910; March 8, 1912; Murray, "Bucket Full," 6.

57. Moore, *Eleventh Annual Report of the Mining Industry of Idaho for the Year 1909*, 86; *Idaho Daily Statesman* (Boise), February 17, 1910.

58. Murray, "Bucket Full," 6.

59. Ibid., 5–6; *Idaho Daily Statesman* (Boise), March 3, 1912; *Mining and Scientific Press* 105 (September 14, 1912): 330; F. Cushing Moore, *Twelfth Annual Report of the Mining Industry of Idaho for the Year 1910* (State Mine Inspector's Report), 7.

60. Robert N. Bell, *Thirteenth Annual Report of the Mining Industry of Idaho for the Year 1911* (State Mine Inspector's Report), 20; *Mining and Scientific Press* 105 (September 14, 1912): 330, 332; *Idaho Daily Statesman* (Boise), March 8, April 1, August 15, 1912.

61. Robert N. Bell, *Fifteenth Annual Report of the Mining Industry of Idaho for the Year 1913* (State Mine Inspector's Report), 165; *Mineral Resources, 1913* (US Geological Survey), 770; *Mineral Resources, 1914* (US Geological Survey), 616–17; *Mineral Resources, 1915* (US Geological Survey), 539; *Mineral Resources, 1916* (US Geological Survey), 170; Etheredge Walker to Charles Janin, Boise, March 4, 1915, Janin MSS, Huntington Library, San Marino, CA (hereafter Janin MSS).

62. *Mining and Engineering World* 44 (January 1, 1916): 35; Ballard, "Geology and Gold Resources of Boise Basin," 31.

63. *Ogden (UT) Standard*, August 22, 1916; *Idaho Daily Statesman* (Boise), August 22, 1916.

64. *Idaho Daily Statesman* (Boise), June 15, 1916.

65. *Mining and Scientific Press* 107 (October 11, 1913): 590; Metzger, "Reconnaissance of Placer Mining," 5. O. H. Metzger says that the big dredge closed down about 1917, which is true. Less accurate is his description of what supposedly happened three years later, when a caretaker shoveling snow from the roof built a fire to keep warm but failed to extinguish it and burned the dredge to the ground. Perhaps it was the five-foot Risdon that burned.

66. *Idaho Daily Statesman* (Boise), December 31, 1916; December 31, 1917; November 3, 1918; *Mineral Resources, 1917* (US Geological Survey), 473; *Eighteenth Annual Report of the Mining Industry of Idaho for the Year 1916* (State Mine Inspector's Report), 87; Stewart Campbell, *Twenty-Fourth Annual Report of the Mining Industry of Idaho*

for the Year 1922 (State Mine Inspector's Report), 45; Frederick S. Mead, ed., *Harvard's Military Record in the World War* (Boston: Harvard Alumni Association, 1921), 307.

67. *Mineral Resources, 1916*, 581; *Eighteenth Annual Report of the Mining Industry of Idaho for the Year 1916*, 36; Robert N. Bell, *Twentieth Annual Report of the Mining Industry of Idaho for the Year 1918* (State Mine Inspector's Report), 79; Campbell, *Twenty-Fourth Annual Report of the Mining Industry of Idaho for the Year 1922*, 45.

68. *Salt Lake Mining Review* 23 (March 30, 1922): 19.

69. *Idaho Daily Statesman* (Boise), April 30, May 15, 1922.

70. *Mineral Resources, 1926* (US Bureau of Mines), 435.

71. *Eighteenth Annual Report of the Mining Industry of Idaho for the Year 1916*, 37.

72. Ballard, "Geology and Gold Resources of Boise Basin," 32; *Transactions of the American Institute of Mining Engineers* (New York: American Institute of Mining Engineers, 1902), 32:lxxi; *Annual Catalogue of the Colorado School of Mines, 1904–1905* (n.p.: n.d.), 108; Murray, "Bucket Full," 5; *Engineering and Mining Journal* 78 (August 25, 1904): 297.

73. George P. Hurst handwritten notebook, Box 4, Hurst MSS; Ballard, "Geology and Gold Resources of Boise Basin," 32; Report on the Grimes Creek Placers, Centerville, ID, December 21, 1912, typed copy, Morris MSS; *Engineering and Mining Journal* 78 (August 25, 1904): 287; Murray, "Bucket Full," 5; *Idaho Daily Statesman* (Boise), March 29, May 21, 1905.

74. *Salt Lake Mining Review* 7 (August 30, 1905): 19; *Idaho Daily Statesman* (Boise), June 8, 1906; *Mineral Resources, 1905* (US Geological Survey), 229; Bell, *Eighth Annual Report of the Mining Industry of Idaho for the Year 1906*, 38.

75. Murray, "Bucket Full," 5; *Mining and Scientific Press* 100 (January 1, 1910): 69.

76. *Idaho Daily Statesman* (Boise), August 25, 1910; *Deseret News* (Salt Lake City), August 29, 1910; Ballard, "Geology and Gold Resources of Boise Basin," 31–33; Robert N. Bell, *Fourteenth Annual Report of the Mining Industry of Idaho for the Year 1912* (State Mine Inspector's Report), 65.

77. Smith was president of the company. Directors were M. A. Regan, E. J. Phelps, Albert Larsen, and M. E. Smith. *Idaho Weekly Statesman* (Boise), August 15, 1912.

78. *Report of Proceedings of the American Mining Congress Eighteenth Annual Session* (Washington, DC: American Mining Congress, 1916), 33.

79. *Mining and Engineering World* 44 (January 1, 1916): 35.

80. *American Mining Manual (Twenty-Eighth Year) 1920* (Chicago: Mining Manual Company, 1920), 186; Campbell, *Twenty-Fourth Annual Report of the Mining Industry of Idaho for the Year 1922*, 48; Murray, "Bucket Full," 7.

81. *Salt Lake Mining Review* 28 (August 30, 1926): 7; *Engineering and Mining Journal* 122 (September 11, 1926): 422; *Who's Who in Engineering, 1922–23* (Brooklyn, NY:

Lewis Historical Company, 1922), 669; Robert S. Crawford, ed., *The Wisconsin Alumni Magazine* 27 (Madison: University of Wisconsin, June 1926): 257.

82. *Salt Lake Mining Review* 28 (August 30, 1926): 7; *Mineral Resources, 1926*, 435; *Mineral Resources, 1928* (US Bureau of Mines), 664.

83. *Mineral Resources, 1926*, 417; *Mineral Resources, 1927* (US Bureau of Mines), 575; *Mineral Resources, 1928* (US Bureau of Mines), 664; *Mineral Resources, 1929* (US Bureau of Mines), 386; *Mineral Resources, 1930* (US Bureau of Mines), 634.

84. *"Boise Payette Lumber Company v. Idaho Gold Dredging Corporation," Idaho Reports* 56 (May 4, 1936): 664–65, 693, 696; *Grant County Journal* (Ephrata, WA), May 14, 1933, clipping, Kimble Database, WSU (hereafter Kimble Database, WSU).

85. *Mineral Resources, 1931* (US Bureau of Mines), 453–54; *Mining Journal*, August 13, 1931, quoted in *Nevada Nugget Hunters*, http://nevadanuggethunters.myfreeforum.org/index.php?component=content&topicid=452&postdays=0&postorder=asc&start=10; *Minerals Yearbook, 1931–32* (US Geological Survey), 101.

86. *Engineering and Mining Journal* 86 (December 26, 1908): 1275; *Engineering and Mining Journal* 123 (April 9, 1927): 617; *Spokesman-Review* (Spokane), August 7, 1927, clipping, WSU Digital Collections, Washington State University, Pullman; Clark Spence, *Northern Gold Fleet* (Urbana: University of Illinois Press, 1996), 29–30, 33; Frank Clinton Merritt, *History of Alameda County California* (Chicago: S. J. Clarke, 1928), 3:399–400; Stewart Campbell, *Annual Report of the Mining Industry of Idaho for the Year 1932* (State Mine Inspector's Report), 97; *Salt Lake Mining Review* 29 (December 15, 1927): 15.

87. *Mineral Resources, 1928*, 663; *Mineral Resources, 1929*, 385; *Mineral Resources, 1930*, 633; Murray, "Bucket Full," 7–8.

88. Oscar C. Perry to Fred Morris, Oroville, CA, December 21, 1912, Morris MSS.

89. J. St. George Joyce, ed., *Story of Philadelphia* (Philadelphia: City of Philadelphia, 1919), 480; Murray, "Bucket Full," 8; *Minerals Yearbook, 1935* (US Geological Survey), 141.

90. William I. Zoch to Charles Janin, Fairfield, ID, February 27, 1937, Janin MSS; *Minerals Yearbook, 1935*, 207; *Minerals Yearbook, 1937* (US Geological Survey), 308.

91. Estimate Proposal, Yuba Manufacturing Company to A. F. Baumhoff and E. T. Fisher of Warren, ID, San Francisco, December 11, 1934, copy, Yuba Manufacturing Company Papers, Meriam Library, Chico, CA.

92. A List of Dredges Built by the Yuba Manufacturing Company, n.d., typescript copy, Charles Romanowitz Papers, Rasmuson Library, University of Alaska Fairbanks; Murray, "Bucket Full," 8.

93. *Mining and Engineering World* 43 (October 2, 1915): 528; *Minerals Yearbook, 1935*, 207–8; *Spokane Chronicle*, January 22, 1936, clipping, Kimble Database, WSU; Murray, "Bucket Full," 8.

94. Metzger, "Reconnaissance of Placer Mining," 29; Murray, "Bucket Full," 9; *Idaho Daily Statesman* (Boise), September 1937; November 27, 2011.

95. *Spokane Chronicle*, June 9, 1937, clipping, Kimble Database, WSU.

96. *Spokesman-Review* (Spokane), July 31, 1935; *Spokane Chronicle*, May 6, 1938, both clippings, Kimble Database, WSU; Murray, "Bucket Full," 8.

97. *Kellogg (ID) Evening News*, October 14, 1937, clipping, Kimble Database, WSU; Metzger, "Reconnaissance of Placer Mining," 31–32.

98. *Minerals Yearbook, 1941* (US Geological Survey), 347.

99. *Minerals Yearbook, 1937*, 312; *Minerals Yearbook, 1938* (US Geological Survey), 340.

100. Metzger, "Reconnaissance of Placer Mining," 25; Murray, "Bucket Full," 8–9.

101. *Minerals Yearbook, 1937*, 320; Murray, "Bucket Full," 10.

102. *Minerals Yearbook, 1937*, 320; *Minerals Yearbook, 1938*, 337; *Minerals Yearbook, 1939* (US Geological Survey), 324; *Minerals Yearbook, 1940*, (US Geological Survey) 327, 340; *Minerals Yearbook, 1941*, 346; Murray, "Bucket Full," 12.

103. "Dredges Designed and Built by Yuba," *Engineering and Mining Journal* 141 (May 1940): advertising section, 29; Edward A. Foster to W. B. Macauley, June 6, 1941, draft copy, Yuba Manufacturing Company Papers, Meriam Library, Chico, CA.

104. Murray, "Bucket Full," 11–12.

105. "Behind Boise," https://sites.google.com/site/behindboise/home/choose-your-tour/harrison-boulevard/1500-Harrison-boulev.

106. Idaho Bureau of Mines and Geology Special Report 1, Idaho Geological Survey, Moscow (1964): 90; *Idaho Blue Book*, "Legislative Branch," n.d., 208, http://www.sos.idaho.gov/elect/bluebook/2012/04_legislative.pdf; Robert Frohlich, "California Dreamin," March 13, 2007, http://www.moonshineink.com/mountain-life/california-dreamin; "Historic Strawberry Lodge," October 2000, https://www.edcgov.us/Living/Stories/Historic_Strawberry_Lodge.aspx.

107. H. S. Lorain and O. H. Metzger, "Reconnaissance of Placer Mining Districts in Idaho County, Idaho," *US Bureau of Mines Information Circular* 7023 (June 1938): 76; Mary Alfreda Elsensohn, *Pioneer Days in Idaho County* (Caldwell, ID: Caxton, 1947), 2:30.

108. *Minerals Yearbook, 1938*, 351; *Minerals Yearbook, 1939*, 325; *Minerals Yearbook, 1940*, 345; *Idaho County Free Press*, June 30, 1976.

109. *Minerals Yearbook, 1939*, 330; *Idaho County Free Press*, June 30, 1976.

110. *Minerals Yearbook, 1941*, 352; *Minerals Yearbook, 1947* (US Geological Survey), 1396; *Minerals Yearbook, 1949* (US Geological Survey), 1475.

111. *Engineering and Mining Journal* 143 (September 1942): 92.

112. *Minerals Yearbook, 1945* (US Geological Survey), 357, 368; *Minerals Yearbook, 1946*, (US Geological Survey), 1427, 1436–37; Murray, "Bucket Full," 13.

113. *Minerals Yearbook, 1947*, 1406; *Minerals Yearbook, 1948* (US Geological Survey), 1499.

114. *Minerals Yearbook, 1949*, 1469; *Minerals Yearbook Area Reports, 1952* (US Geological Survey), 287.

115. *Minerals Yearbook, 1949*, 1469; *Minerals Yearbook, 1950* (US Geological Survey), 1491; *Minerals Yearbook, 1951* (US Geological Survey), 1489; *Minerals Yearbook Area Reports, 1952*, 287; Murray, "Bucket Full," 14.

116. See, for example, Richard M. Patterson, *Historical Atlas of the Outlaw West* (Boulder: Johnson, 1988), 52.

117. Idaho State Historical Society Reference Series 9, "Mining in Idaho," Idaho State Historical Society, Boise (revised 1985), 9; A. H. Koschman and M. H. Bergendahl, "Principal Gold-Producing Districts of the United States," Geological Survey Professional Paper 610 (Washington, DC: US Government Printing Office, 1968), 125.

118. *Minerals Yearbook, 1951*, 630; *Minerals Yearbook, 1957* (US Geological Survey), 359.

13

Dredging Rare Metals

Ruby Meadows is a comparatively shallow basin of approximately 1,000 acres, located twelve miles west of Warren, four miles south of Burgdorf, and thirty-two miles northeast of McCall, the nearest railroad terminal. Gold was first discovered in 1863, and a small mining settlement grew up at Miller Camp and Larson Diggins on a ridge just above Ruby Creek. The camps were overshadowed by the Boise Basin boom and never prospered to any great extent. In the twentieth century, the creek itself was not mined until 1932, when Jack West built a small cabin and, after hard work, took out $300 in gold. Five years later, higher technology came to Ruby Meadows in the form of a small, improvised dragline and washing plant on a stretch of ground running a few hundred feet above the mouth of Miller Gulch. This was the jerry-built old steam outfit from Sand Creek, moved by Frank Jones and sons to its new location under the name Ruby Creek Dredging Company. The cut was about fifty feet, with an average depth of four feet to decomposed granite bedrock. Roughly $1,500 in gold was taken out, but the sluices also contained large quantities of monazite and cinnabar concentrates, which were

DOI: 10.5876/9781607324751.c013

not what the quest was about. The 1937 run was short; after only five weeks Jones it shut down because values decreased as digging proceeded.[1]

By the late 1940s, priorities for dredge mining were changing. Large blocks of suitable gold placer ground were becoming increasingly difficult to find, and a new emphasis was placed on strategic heavy metals, including many of old black sands, which now became valuable. After World War II, more than one dredge owner diverted his own rig or found a new one and stepped up the chase for zircon, monazite, ilmenite, garnet, or some other industrial mineral.

Rare Earths, Inc., was one example. The firm was quick to recognize the increasing values of strategic minerals and to make at least one weak attempt to mine them at Ruby Meadows. Organized in the spring of 1948, the firm spent most of the next nine months preparing before it issued its prospectus. Up to that time, all funds for development, exploration, machinery, labor, and other costs had been provided by L. Waggoner, B. Munson Waggoner, and A. C. Solomon, plus a loan from the Lindsay Light and Chemical Company (LL&C). The LL&C originally focused on manufacturing incandescent gas mantles using thorium nitrate to make the light glow more brightly. After LL&C's supply of German thorium nitrate was cut off during World War I, it began refining monazite sand to extract thorium in west Chicago. When electric lights replaced gas, the firm gave thorium production top priority and in the 1940s was a major provider for the Manhattan Project and the atomic bomb. After World War II, LL&C continued to refine rare-earth metals for the Atomic Energy Commission—which explains the firm's willingness to underwrite a loan for a small untested company in faraway central Idaho.[2]

The prospectus of Rare Earths, Inc., was in some ways typical of the genre and in other ways distinctive. True, it was a sales document, designed to conjure up investors, but it was much more modest than most. If it spoke of Ruby Meadows's "long history of prospecting and intermittent hydraulic mining on a small scale," it pointed out that the Meadows proper had never really been mined, although it had been tested several times for gold—the last time in 1933 by Walter Hovey Hill. To Hill's fifteen-year-old estimates of dredge-able ground available to the company, the prospectus added another 100,000 cubic yards for the additional claim, Monazite Placers, which had been staked since Hill's examination to make a total of 6,270,400 cubic yards. In 1933 Hill had calculated gold values of 31 cents a yard on the basis of an average of twenty-two samples, using the metal's old price. Lindsay Light and Chemical Company also made a crude analysis of the gravel for heavy metal concentrates in three tests, showing an average value for monazite of 0.33¾ cents a cubic yard and that of garnet at $2.40 per yard. That said, the prospectus emphasized several times that exact evaluations "are not available in a sufficient number of tests to make definite and accurate statements as to the precise dollars and cents which can be recovered." Like most such documents, the prospectus promised a second dredge, if favorable conditions warranted.[3]

Rare Earths, Inc.'s, ground on Ruby Meadows was leased for a ten-year period, with an option to purchase from a number of people living in the McCall area and Idaho City. Royalties to that group were to be 13 percent of gross receipts for all minerals marketed. For machinery, the firm contracted with the Wallberg Dredging Company, an Idaho corporation with California investors, to work the property. Ivor O. Wallberg, head of the dredge firm, had patented a gravel pump and designed his own suction dredge with a capacity of 100 yards per hour. A seasoned mining man, Wallberg had been active in that field earlier in California, and in March 1948 his machine was digging monazite and zircons on the Salmon River somewhere near White Bird.[4] During the summer it was dismantled, moved to Ruby Meadows, and reconstructed, with an additional pontoon added to provide space for specially designed concentration tables necessary to handle heavy mineral concentrates. Under the contract, Wallberg Dredging would pay all dredging costs and half the expense of transporting the concentrates to the separation plant in McCall. In a separate agreement with Hart and Oberg, hauling contractors, the price of trucking these products was set at eight cents per ton per mile. In return, the Wallberg Dredging Company would receive half of the gross receipts of all products it recovered, after the royalty payments to the lessors had been met. Rare Earths' spokesmen approached the Idaho Garnet Abrasive Company of Spokane, which indicated that it would be glad to sell the dredge's entire output of garnet through its brokerage firm at prices ranging from forty dollars to sixty-five dollars per ton, but no commitments were made.[5]

When snow closed the roads into Ruby Meadows in the fall of 1948, the dredge had been operational for only three weeks, and much of that time had been devoted to making various adjustments. Supplies were cached on the dredge site, and it was planned that as soon as the weather permitted in the spring, Ivor Wallberg and his dredgemaster, George Guiler, would fly to Burgdorf, then ski or snowshoe in to prepare the machine for work.[6] Whether that actually happened is dubious. If it did, the dredge did not likely operate for long.

In 1950, engineer Walter Hovey Hill wrote another report on Ruby Meadows for a different client and tried to calculate the production of the Wallberg suction dredge. "As far as can be determined," he noted, it had worked 8,755 cubic yards, on an area stripped previously by a dragline bucket. About $2,000 in gold had been recovered, and 3 tons of monazite had been processed in McCall and were ready for shipment to LL&C in Chicago. Also, 30 sacked tons of garnets were awaiting disposition. Hill had disparaging remarks for the "promiscuous sampling of the basin for monazite by Rare Earths, Inc.," and was "of the opinion that . . . the values in gold and magnesite were [not] recovered in the crude Walburg Dredge."[7]

During the Cold War years, beginning a few years after the end of World War II, the federal government's concern about the shrinking stockpile of strategic minerals from abroad peaked as a result of embargos on magnesite by Brazil and India. This led to a two-pronged campaign to build reserves, not only of this particular mineral but of a wide range of others as well. One approach was to encourage exports from elsewhere in the world by paying higher prices for the required materials. The other was domestic—to create several programs designed to further the exploration and discovery of certain strategic and critical minerals. This included the so-called rare-earth metals—the fifteen chemical elements in the periodic table from lanthanum, no. 57, through lutetium, no. 71, plus two others, scandium and yttrium, with similar properties. All of these minerals had become increasingly important to modern technology and would become even more so. As time passed, color TVs, automotive catalytic converters, battery electrodes, lasers, X-ray tubes, electronic equipment, sophisticated military weapons, and countless other products came to depend on one or another of these metals. Also on the "must" list were better-known metals such as lead, copper, zinc, manganese, uranium, vanadium, and a number of others, eventually including gold and silver.[8]

The Defense Production Act of 1950 provided financial aid to private enterprises for the production of goods and services necessary for national security. Under the law's provisions, a few months later the secretary of the interior established the Defense Minerals Administration (DMA) to work as a direct contracting agency for the exploration and discovery of key minerals. The DMA program was administered from Washington, DC, by a small group of veteran mining engineers and geologists drawn from the business world. Under their guidance, the DMA laid down ground rules under which an individual or a company might apply for financial assistance to further explore a property, owned or leased. Field teams, made up of US Geological Survey specialists and US Bureau of Mines engineers, examined the proposed site and, if accepted, followed through with subsequent testing and reports. In the relatively small number of cases in which significant discoveries were made, a certificate of possible production might be issued to the operator, detailing the payment of royalties to the government should actual mining proceed.[9]

The Defense Minerals Administration lasted less than a year. It was terminated on November 21, 1951, and transferred to the Defense Minerals Exploration Administration (DMEA) program, which now added mineral exploration to its other agendas. Administered by the same personnel from the same offices as the DMA, the DMEA was committed to the same mission—the discovery of critical and strategic minerals. Some DMA contracts were completed by the new agency; others were rewritten. The program was greatly expanded and it continued until 1958, when government support was no longer considered justifiable, since the defense needs of mineral supplies were considered to have been met. But the door was left open with the creation of a new Office of Minerals Exploration (OME), which was similar to previous DMA and DMEA programs but more restrictive.[10]

To reduce costs, administrative and operating responsibilities of the OME program were transferred to the US Geological Survey in 1965. Funds allocated to the survey were terminated in 1974, with limited appropriations for four more years to close out projects already under way. When further changes in handling federal aid for exploration became effective in January 1993, the OME was closed out completely.[11]

In his report on all these programs, Thor Kiisgaard lists twenty pages of "Dockets," at least 350 files of applications made for support in Idaho under all three programs. Most were under the DMEA, and some individuals or companies made more than one application. Many were made for exploration of lead, copper, zinc, or other metals ordinarily found in lode mines. Far fewer, perhaps 15 percent, were for the rare or heavy minerals that could be dredged, and the degrees of success in that category were small.[12] Many were called, but few were chosen.

Some of those who fell by the wayside had their applications rejected from the beginning because of a lack of information. Others followed through with various stages of their project, only to be disappointed when their ground proved unprofitable, as in the case of the Cosumnes Gold Dredging Company, a California firm with holdings in Paddy Flat, northeast of Lowman in White Hawk Valley. Cosumnes actually had two projects in the same area but dropped them both before completing the final stage.[13] Still others were fortunate enough to satisfy all government requirements and actually put a dredge on the ground.

Among those in that final category was the Ruby Meadows Mining Company, with leased property in the Burgdorf area and headed by George C. Kosmos. Its brief history provides an interesting demonstration of a sputtering new dredge enterprise under mediocre leadership, as well as the role of a bureaucratic federal government agency. Kosmos had appeared on the Idaho mining scene earlier, in late 1947, as part of a Seattle company under the leadership of B. M. Sites and Co., which announced that it had bought a dredge at Sulphur, Nevada, and would move it to Ruby Meadows and operate on land owned by Dr. C. C. Fairchild and associates of Idaho City.[14] There is no indication of the type of dredge or if this plan ever came about. In all likelihood, it did not.

By the early 1950s Kosmos was president and manager of the new Ruby Meadows Mining Company, in a limited partnership that included Kenneth J. Selander and M. L. Davies, both, like Kosmos, from Seattle. In March 1951, ignoring formal application requirements, Kosmos submitted a request for a $420,000 loan from the Defense Minerals Administration to provide working capital for a magnesite dredging operation on Ruby Meadows. He was informed that the ground involved had not been adequately tested as required by agency rules and that he should apply for an Aid for an Exploration Project. Included were five copies of Form MF–103, to be filled out in quadruplicate.[15]

In his application for money to determine the value of magnesite, zircon, and other minerals of his firm's ground and also on Ruby Meadows, Kosmos described himself as having "been in the construction and mining business in California, Alaska, and Idaho since 1910." He had worked for the Kennicott Copper Mining Company and contracted for building part of the Alaska Railroad. He also noted that in 1937 a small dragline had been operated by J. T. Jones on part of the property, and the Wallburg suction dredge had worked 8,755 cubic yards and taken out about $2,000 in gold, 3 tons of magnesite, and 30 tons of garnets. There was no current activity, he said, but the Yuba Dredging Company was installing a dredge on the lower Secesh River, six miles from Ruby Meadows.[16]

The ground itself consisted of 800 acres in six unpatented placer claims under lease from Kosmos, who in turn had acquired leases and options in 1946 from the diverse owners, including E. M. Stites and Dr. and Mrs. C. C. Fairchild, who had been involved in Kosmos's previous Ruby Meadows venture. The applicant had arranged for William A. Noon, a prominent Montana engineer and dredge manager, to oversee the systematic drilling and sampling, which would be completed within one month at a cost of not more than the $14,000 he was asking from the Defense Minerals Administration. Kosmos was careful to note that if he were successful, he would be back for more funds, and he emphasized that he had included an estimate requested by a colleague from Washington Iron Works. A tentative price for a steel hull four-and-a-half-foot bucket dredge was $275,000. Add to that another $60,000 for a 250 kilowatt diesel electric-generating plant with underwater cables, plus an additional $25,000 for whatever special jigs were required for handling strategic metals—everything FOB from Seattle. A spokesman for Washington Iron Works added that a used duplicate of that machine in excellent condition was being offered for sale in Fort Jones, California, by the French Gulch Dredging Company.[17]

The company completed its testing, but E. J. Carlson, the US Bureau of Mines engineer attached to the Special Minerals Investigation Branch, believed the sampling had been sub-standard and that more would be needed to get an accurate evaluation of reserves. Kosmos did not disagree but indicated that he was satisfied with the results and intended to put a dredge in operation in the spring of 1952 as a private financial venture.[18] When finished, crews had churn-drilled fifty-one holes to an average depth of 22.9 feet, with total estimated yardage of 4.58 million cubic yards. By Kosmos's reckoning, each yard of gravel contained 2.18 pounds of magnesite and 10 cents worth of gold, and there were 2 million to 4 million additional yards of workable ground in other creeks adjacent to the property already explored. Kosmos also estimated the gross recoverable value per cubic yard at $0.465 and overall costs of dredging at $0.1436 per yard at a rate of 70,000 yards per month. Some critics seemed to question the accuracy of the sampling. There were hints of deliberate or accidental salting, but cooler heads prevailed, arguing that while not perfect, testing had been within the reasonable limits of placer examination. The drill foreman and panner were veterans with solid reputations for excellent work, who had left many successful mining enterprises in their wake. When others challenged the estimated cost of dredging, it was pointed out that the cost was based on William Noon's actual expenses while managing the boat of H and H Mines, Inc., on Henderson Creek in Montana from April 1 to December 20, 1949—the very dredge Kosmos was intending to use at Ruby Meadows. In the end, the DMEA awarded the company a "Certificate of Discovery," which entitled it to reimbursement of 90 percent of its $14,000 in expenses to that point but also committed the firm to pay a 5 percent royalty to the government on any values dug for a ten-year period.[19]

In March 1952 Kosmos again asked the DMA for a loan, this time to produce magnesite concentrates. The request was for $230,000 "to procure and install a bucket-line dredge, living quarters, shops, bulldozer, and working capital." The field team believed the loan was justified and so indicated to agency officials in Spokane, who forwarded their views to Washington.[20] No evidence indicates that it was approved at the higher level.

Through private financing, the Ruby Meadows Mining Company did eventually manage to field a dredge, not in 1952, as Kosmos had hoped, but a year later and not under its own name but as the K and D Mining Company, created as a subsidiary. As indicated on his modified government application, Kosmos had hoped to use the H and H Mines, Inc.'s, Montana boat that William Noon had managed before he was appointed a member of Ruby Meadows Mining's DMEA field team. That machine was a three-foot all-electric Yuba, which was dismantled and moved from Crooked River in Idaho in 1942. It was reassembled, equipped with special jigs, and commenced digging both gold and tungsten-valuable scheelite south of Drummond on May 2, 1943. The boat continued to dig throughout World War II to the end of its final season in December 1949.[21]

Although the direct lineage is not clean-cut, this was very likely the K and D dredge that began work about June 1953 under the supervision of a dredgemaster named Smith, who reported three months later that its operation was still below the desired efficiency for profitable production. The recovered gold went to the US Assay Office in Seattle; magnesite and garnets were forwarded to various processors in Weiser and Boise for further separation.[22] The going was not easy. The dredge lost two-and-a-half months of production when it sank during the summer, and shortly thereafter it was forced to shut down during the long winter months. The DMEA's semiannual report for the last half of 1954 showed no activity by the rig since the previous September, and the Forest Service indicated that the idle machine was still on the property in early June, although inaccessible to the examining engineer because of the snowbound access road.[23]

That ended the K and D dredge's brief operations on Ruby Meadows. It was never a paying proposition, and no doubt the falling price of monizite helped send the company under. Its production record was controversial, and Kosmos's bookkeeping turned out to be sloppy. He failed to report the sales of gold, silver, magnesite, and garnets to the DMEA and for several years avoided paying any royalties, forcing the government to trace the minerals through the buyers to determine how much had been sold and at what price. Final figures ascertained that in its short life span the dredge had recovered 241.033 ounces of gold, 50.83 ounces of silver, 108.845 tons of garnet sands, and 10.132 tons of magnesite, for a total net value of $16,609.26. After hounding the company for payment with help from the Ada County sheriff and the US attorney general, the DMEA finally collected its royalties of $830.46 and breathed a sigh of relief.[24]

Kosmos noted in early 1955 that the Ruby Meadows Mining Company had been out of funds for the last year and a half and that its property had been leased to a new entity, K-D Enterprises, which was negotiating with "a large concern" with enough capital to resume operations. A little later he said he had concluded a deal with the International Metal and Oils Company to take over ownership and resume operations.[25] Nothing more was heard of this proposed new enterprise.

George Kosmos was no more successful with a kindred business venture that involved a by-product of his dredging activity. Magnesite, a thorium-bearing rare-earth phosphate, was a slightly radioactive mineral on which the imaginative Seattle-ite set out to capitalize. Through a separate firm, Raco, Inc., he conceived the idea of the Cosmos Radioactive Pad, a small quilted pad filled with two or three pounds of magnesite and touted as a cure for a number of maladies. By the end of January 1955, according E. A. Magill, the DMEA mining engineer, he had sold 300 of his quack therapeutic bags. Soon, the Federal Drug Administration brought sales to a halt with a Notice of Judgment against Raco, Inc., for its misleading claims that its pads "provided an adequate and effective treatment for arthritis, bursitis, rheumatism, neuritis, and sinus trouble, and soreness of hands, wrists, forearms and back."[26]

If Kosmos and the Ruby Meadows Mining Company's experience was hardly a success story, there were others that were. Porter Brothers Corporation, an old-line Spokane construction company, is a shining example. At least three Porter brothers became important figures in the mineral industry of the Intermountain West. Richard D. Porter was the most sedentary of the three and tended to work out of his hometown. Andrew Porter was associated with Portland and eastern Oregon, while Robert, probably the most active, had headquarters in Helena and Boise at different times. They built solid and lengthy reputations for gold dredging in both Montana and Oregon and eventually in Idaho as well. Their six-foot electric Yuba worked in Last Chance Gulch near Helena almost continuously from November 1935 to August 1943, when gold mining was banned during World War II. Resuming work in July 1945, the machine dug with only slight interruptions for another five-and-a-half years before it completed its final run at the end of the 1950 season.[27]

During much of the same period—beginning on September 10, 1938, and operating as Porter and Company—the brothers ran a four-and-a-quarter-foot Yuba bucket dredge on a number of creeks in Grant County, Oregon. After the mandatory wartime closing in 1941 that lasted through 1946, the machine finally wound down for good in the spring of 1951.[28] Thus, with completion of its work in Montana and Oregon somewhat tentative, it is not surprising that the company displayed a serious interest in the new government aid programs for strategic minerals. Having done considerable investigation in advance, Porter Brothers Corporation made a very ambitious application for government aid to the Defense Minerals Administration in May 1951 to drill 115 test holes in Bear Valley, fifteen miles north of Lowman in Valley County. The original request was for $50,000, with work to be completed during the 1953 season, but after early sampling showed tremendous potential for columbium-tantalum–bearing magnesite, with complex and expensive laboratory work to separate the components, the Spokane firm asked that the figure be revised upward to $92,000. Columbium (also known as niobium) was important as a high-temperature alloy for jet engines and gas turbines. Tantalum in pure form was a highly corrosion-resistant metal, useful not only in chemical equipment but also in electronics and optical glass. The US Atomic Energy Commission showed considerable interest in the proposal, and President Robert Porter visited New York to discuss the Bear Valley property and program with Dean Frasche of Union Carbide and Carbon Company, specifically a possible commercial venture should the deposit prove out. Porter explained that if at any point the government did not think the project warranted further investigation, the work would be discontinued and no further cost entailed. "We are dredge operators and are not interested in prospecting alone, but only as a means to develop a dredging project," he said.[29]

For reasons that are not completely clear, Porter Brothers was slow to move toward an active dredging operation. Perhaps bureaucratic red tape hampered the go-ahead agreement, or the delay could have been caused by the need to refit dredges with the special equipment required and possibly the need to build a sophisticated processing plant. Whatever the cause, in the end the company emerged with a government purchase contract to produce 1.05 million pounds of columbium-tantalum concentrates of 90 percent pentoxide content by June 21, 1961. Startup was expected in 1954, but the firm was still erecting dredges on its 11,500-acre deposit, and a processing plant was not completed by the end of the year.[30]

The company began production with one dredge in 1955 and added another early the next year. One was a six-footer, in all likelihood the old Helena rig. The other was a four-and-a quarter-footer, probably the veteran of the Oregon campaigns. Both were equipped with jigs to recover black sands. These dredges were now the only producers of those rare-earth metals in the state. The crude concentrates from the boats were trucked to the firm's plant in Lowman for electromagnetic-electrostatic separation of the various components—where relatively clean concentrates of euxenite and columbite and less clean concentrates of magnetite, ilmenite, garnet, and zircon were separated and the results sent to the Mallinckrodt Chemical Company in St. Louis for further processing. Dredging in Bear Valley came to a close in October 1959, when the Porter Brothers' General Services Administration contract was fulfilled. Between them, the two boats had handled about 6.5 million cubic yards of placer ground and produced 2,049 short tons of euxenite, 54,862 tons of ilmenite, 83.5 tons of columbite, and an unrevealed tonnage of magnesite. The Porters' mill at Lowman and their two dredges in Bear Valley, although idle, were maintained in operating condition for several years, in hopes of renewed government contracts to resume dredging. These contracts depended on suitable markets for the mill products, which never developed.[31]

In the meantime, while the Porter Brothers saga played out and the government was doing its best to encourage production of rare-earth minerals, several established gold-dredging companies had moved to the newer fields on their own, without federal support. In 1952 Gem Magnesite Mines, Inc., was reportedly doing preparatory work in anticipation of dredging in another large area of radioactive material in the southwestern part of Long Valley, in Valley County. Sheridan K. Atkinson, a former dredge operator in the Boise Basin, was cleared by both the Oregon and Idaho State Land Boards to dredge black sands along the Snake River near Weiser, although neither Gem Magnesite nor Atkinson seems to have followed through.[32]

The most successful of the privately induced magnesite dredging efforts occurred in Idaho's Cascade Valley on Big Creek, four miles southeast of the town of Cascade. By 1948, growing desperate in the face of increasing demands, two representatives of LL&C, the largest processor of rare-earth minerals in the county, approached veteran dredge operator Fred Baumhoff to seek new sources of the scarce heavy metals. Baumhoff first looked at the Boise Basin, where he knew from experience that magnesite was present and had been ignored for years. But ground in the basin had been dredged and re-dredged so much that he concluded it would be impossible to work, so he next considered the Cascade Valley, where magnesite had been found not long before. Three companies leased placer property along Big Creek, as did Jack Simplot, the potato king who was also heavily invested in mines. Subsequent drill testing was favorable, although large deposits were spotty and were found only on westward-flowing streams.[33]

In 1950 Baumhoff's own company, Baumhoff-Marshall, Inc., moved its bucket dredge from Centerville and started up in September. During the following summer it was joined by two more boats—one operated by the Idaho-Canadian Dredging Company, also from Boise Basin, and the other owned by the Warren Dredging Company, also a Baumhoff-Marshall firm, which had just finished with the Wild Rose group of claims near Elk City. Simplot would not put a dredge on his lease, but he did provide some of the financial underwriting.[34]

Baumhoff-Marshall's dredge was a six-foot Yuba built to recover gold. With an eighty-five-bucket ladder, it was able to dig thirty-five feet below the water line. Its average capacity for digging magnesite sands was 5,000 cubic yards per day. Like the other two boats, it had to be modified to handle the "blacks," as the sands were called. Its buckets dumped into a six-foot-diameter, forty-eight-foot trommel screen, but the fines flowed directly into one of twelve double Pan-American jigs, six on each side of the trommel, so placed that each set drew sand from a different section of the trommel. The final concentrate was then streamed into a large storage tank in the middle of the main deck underneath the trommel, from whence it was pumped through a two-inch rubber air hose to a dumptruck onshore.[35]

The other two dredges started up in late summer of 1951. The dredge of the Idaho-Canadian Dredging Company was also a six-foot Yuba that had been operated for several years by the Star Pointer Exploration Company at Bearclaw, Montana, then moved to Boise Basin. Its ground was on Big Creek a quarter of a mile west of the Baumhoff boat, and the machine was a little later model than its counterpart and could dig a bit deeper. It was also equipped with more jigs—thirty Pan-American and "Yuba rougher jigs." For the most part, the two dredges were similar, including the method of off-loading the concentrates, except that Idaho-Canadian used a movable screening and de-watering plant onshore before trucking the sands to Cascade.[36]

The third magnesite dredge on Big Creek was a four-foot machine owned by the Warren Dredging Company, which was transported from Warren and started work in late 1951. It dug on ground west of both of the other boats, with apparent success until May 1953, when a high-pressure pump discharge line broke, filling a pontoon with water and capsizing the dredge. The machinery was salvaged, but it was not rebuilt in the Cascade Valley.[37]

All three of these boats worked under the same conditions. Irregular and wide variation of "blacks" in the sand was a problem met by the use of Keystone drill testing ahead of the dredging. Idaho-Canadian's dredgemaster, Miles Young, devised a new system of sluices parallel to the trommel to get a more even jig flow, which helped. Despite this uncertainty, Jack

Fischer, Baumhoff's dredgemaster, estimated the recovery rate at 90 percent. When a rich concentration of magnesite was found, a winchman could readily spot the colors and slow the bucket-line speed accordingly. In the winter, the dredges continued to work even in heavy snowfall and temperatures as low as forty degrees below zero. In such cases, the pond was narrowed to prevent ice from forming between dredge swings. No headlines were used; a single spud provided the pivot. Since all the digging was fine gravel with no boulders and the load on the stacker belt was lighter, the stacker itself was shortened. The Idaho-Canadian dredge had another innovation—a nozzle on the end of the stacker, through which a high-speed jet of water could be pumped to help settle and level the tailings left behind in hopes of more fully meeting state regulator laws just passed for gold dredging.[38]

Concentrates from the dredges were trucked to Cascade and stockpiled separately alongside a spur of the Union Pacific Railroad, from whence they were loaded onto railroad gondola cars by a Bay City clamshell and sent to the Baumhoff-Marshall sand plant in Boise. There, the material was dried, then separated into ilmenite, magnesite, and zircon by electromagnetic and electrostatic equipment. The first two minerals were stockpiled, and the magnesite was sent to the General Services Administration and the Lindsay Chemical Company, the former Chicago-based Lindsay Light and Chemical Company, which had readjusted its focus and changed its name. Idaho-Canadian and Baumhoff-Marshall dredges ceased digging in 1955. Because of falling prices and the lack of a market, it was a short season. Both machines closed down in the summer, remained idle through 1956, and never started up again.[39]

The three Big Creek dredges had provided the country's supply of magnesite during the years they operated, but production statistics were never released. Attracted by high profits driven by temporary scarcity, they contributed to their own undoing. Their heavy output may have helped Lindsay Chemical Company weather a severe storm, just as it helped bolster the government stockpile of magnesite. At the same time, the dredges, combined with the federal program to encourage the discovery and recovery of the rare metals not only in the United States but elsewhere in the world, could only drive down market prices. Ultimately, in the twenty-first century, after experiencing several generations of cheap imported minerals, Americans would wake up to the realization that 95 percent of their heavy metals were coming from China. In 2010, when that nation announced its intention to sharply cut back the export of these commodities, the scramble was on to rediscover and exploit those resources in the United States.

Notes

1. Walter Hovey Hill, *Report and History on the Ruby Gold Placers in Warren Mining District, Idaho County, Idaho* (n.p.: September 27, 1950), copy, Ruby Meadows Mining Company Docket 0930, http://minerals.usgs.gov/dockets/id.htm (hereafter RMMC Docket 0930); *Lewiston (Idaho) Mining Tribune,* October 5, 1936, clipping, Kimble Digital Collections, WSU; H. S. Lorain and O. H. Metzger, "Reconnaissance of Placer Mining Districts in Idaho County, Idaho," *US Bureau of Mines Information Circular* 7023 (June 1938): 64.

2. "Lindsay Chemical Company," n.d., Lehman Brothers Collection, Harvard Business School, Baker Library Historical Collections, http://www.library.hbs.edu/hc/lehman/chrono.html?company=lindsay_chemical_company.

3. Rare Earths, Inc., *Prospectus* (December 1948), 18, 20–23, copy, RMMC Docket 0930.

4. Ibid., 20; Ivor O. Wallberg, Gravel Pump, patent no. 2,281,733, May 5, 1942, Google Patents; W. W. Staley, "Distribution of Heavy Minerals in Idaho," Idaho Bureau of Mines and Geology Mineral Research Report 5 (Moscow: Idaho Bureau of Mines and Geology, March 1948), 6.

5. Rare Earths, Inc., *Prospectus*, 19–21.

6. Ibid., 21.

7. Hill, *Report and History on the Ruby Gold Placers*.

8. Keith R. Long, Bradley S. Van Goosed, Nora K. Foley, and Daniel Corder, "The Principal Rare Earth Elements Deposits in the United States—a Summary of Domestic Deposits and a Global Perspective," US Geological Survey, *Scientific Investigations Report 2010–5220* (n.p., 2010), 3; M. H. Stats, R. B. Hall, D. L. Make, T. J. Armbrustmacher, and I. K. Brownfield, "Thorium Resources of Selected Regions in the United States," US Geological Survey Circular 824 (1980): 10; Thor H. Kiilsgaard, "Mining Properties in Idaho That Were Involved in the DMA, DMEA, or OME Mineral Exploration Programs, 1950–1974," US Geological Survey, Open File Report no. 97-429 (Spokane: US Geological Survey, 1997), 1–3.

9. Kiilsgaard, "Mining Properties in Idaho," 11–34.

10. Ibid., 3–4.

11. Ibid., 4.

12. Ibid., 11–34.

13. Merrill Yost to US Bureau of Mines, Red Mountain, CA, June 7, 1953, copy; John D. Bardill and J. Hoover Maclain, "Final Report," Spokane, July 1954, both in Cosumnes Gold Dredging Company Docket 2265, http://minerals.usgs.gov/dockets/id.htm.

14. *Engineering and Mining Journal* 148 (December 1947): 107.

15. E. A. Magill, Memorandum to DMEA Field Team, Region 1, Spokane, June 6, 1955), copy; Tom Lyon to George C. Kosmos, Washington, DC, May 8, 1951, copy, both in RMMC Docket 0930.

16. George C. Kosmos to Defense Mineral Administration, cover letter and application, Seattle, May 15, 1951, copy, RMMC Docket 0930.

17. Ibid.

18. A. E. Weissenborn and M. E. Volin, Memorandum to Director, Production Expansion Division, Defense Minerals Administration, Spokane, October 18, 1951, copy, RMMC Docket 0930.

19. J. D. Bardill, Memorandum to M. E. Volin, April 30, 1952, copy; A. E. Weissenborn and M. E. Volin to Frank E. Johnson, Spokane, May 8, 1952, copy; C. O. Mittendorf to Jess Larson, Washington, DC, May 14, 1952, copy; all in RMMC Docket 0930.

20. A. E. Weissenborn and M. E. Volin, Memorandum to Chairman, Operating Committee, DMEA, Washington, DC, and Spokane, WA, March 21, 1952, copy, RMMC Docket 0930.

21. J. D. Bardill, Memorandum to M. E. Volin, Spokane, April 30, 1952, copy, RMMC Docket 930; *Minerals Yearbook, 1943* (US Geological Survey), 399; *Minerals Yearbook, 1949* (US Geological Survey), 1237. See also *Engineering and Mining Journal* 150 (June 1949): 119.

22. John A. Bowsher, Memorandum to DMEA Field Team, Region II, Spokane, September 23, 1953, copy, RMMC Docket 930; *Minerals Yearbook Area Reports, 1953* (US Geological Survey), 330.

23. E. A. McGill, Memorandum to DMEA Field Team, Region II, Spokane, January 29, 1954, copy; J. D. Bardill, Semi-Annual Report of Certified Project DMA-930X, July 1, 1954, to December 31, 1954, copy, both in RMMC Docket 930.

24. E. A. Magill, Memorandums to DMEA Field Officers, Region I, Spokane, March 2, April 4, August 8, 1955; D. R. Maclaren to Sheriff, Ada County, Spokane, August 9, 1962; J. R. Armstrong to Herbert Brownell Jr., Washington, DC, January 30, 1956; George Cochran Doub, Assistant Attorney General, to Solicitor, Department of the Interior, Washington, DC, February 12, 1957, all copies, RMMC Docket 0930.

25. George C. Kosmos to E. A. Magill, McCall, ID, January 25, 1955; E. A. Magill, Memorandum to DMEA Field Team, Region II, Spokane, January 10, 1955; E. A.

Magill, Memorandum to DMEA Field Team, Region I, Spokane, March 9, 1955, all copies, RMMC Docket 0930.

26. E. A Magill, Memorandum to DMA Field Team, Region I, January 28, 1955, copy, RMMC Docket 0930; "Cosmos Radioactive Pad (ca. 1950–1955)," http://www.orau.org/ptp/collection/quackcures/cosmos.htm.

27. Charles J. Lyden, *The Gold Placers of Montana*, Montana Bureau of Mines and Geology Memoir no. 26 (Butte: Montana Bureau of Mines and Geology, 1948), 56–57; *Spokane Chronicle*, July 30, 1935, clipping, Kimball Digital Collections, WSU; *Engineering and Mining Journal* 139 (February 1938): 88; *Engineering and Mining Journal* 141 (December 1940): 80; *Engineering and Mining Journal* 146 (September 1945): 120; *Engineering and Mining Journal* 151 (November 1950): 133; *Minerals Yearbook, 1950* (US Geological Survey), 1530.

28. *Minerals Yearbook, 1938* (US Geological Survey), 447; *Minerals Yearbook, 1942* (US Geological Survey), 468; *Minerals Yearbook, 1946* (US Geological Survey), 1518; *Minerals Yearbook, 1951* (US Geological Survey), 1583. Although the annual *Minerals Yearbook* throughout the period of the boat's work in Oregon insisted that it was a four-and-a-half-cubic-foot machine, two of the brothers and the Yuba manufacture list are accurate in calling it of four-and-a-quarter-foot capacity. "List of Dredges Built by Yuba Manufacturing Company," n.d., Charles Romanowitz Papers, Elmer E. Rasmuson Library, University of Alaska Fairbanks; R. Porter, R. B. Porter, and R. A. Lathrop, "Industrial Minerals—Dredging within the Law," *Mining Engineering* 12 (May 1960): 468.

29. *Minerals Yearbook, 1951*, 1353; Defense Minerals Administration, Review of Application for Exploration Assistance, Porter Brothers Corporation, July 5, 1951, copy; Robert Porter to Weston G. Bourret, Spokane, November 9, 1951, copy, both in Porter Brothers Corporation Docket 1899.

30. *Minerals Yearbook Metals and Minerals (Except Fuels), 1954* (US Geological Survey), 1:115; *Mining World* (San Francisco) 16 (September 1954): 109.

31. M. H. Stats et al., "Thorium Resources of Selected Regions in the United States," 9–10; *Minerals Yearbook Area Reports, 1955* (US Geological Survey), 3:365; *Minerals Yearbook Area Reports, 1956* (US Geological Survey), 3:364; *Minerals Yearbook Area Reports, 1962* (US Geological Survey), 3:345; *Minerals Yearbook Area Reports, 1964* (US Geological Survey), 3:325.

32. *Minerals Yearbook Area Reports, 1952* (US Geological Survey), 3:287.

33. George Argall Jr., "New Dredging Techniques Recover Idaho Magnesite," *Mining World* (San Francisco) 16 (February 1954): 26–27.

34. Ibid., 26–27, 30; *Minerals Yearbook Area Reports, 1953*, 3:278, 305.

35. Argall, "New Dredging Techniques Recover Idaho Magnesite," 28.

36. Ibid., 28–29.

37. Ibid., 29; *Minerals Yearbook Area Reports 1953*, 3:330, 350.

38. Argall, "New Dredging Techniques Recover Idaho Magnesite," 28–29.

39. Ibid., 30; *Minerals Yearbook Area Reports 1955*, 3:365; *Minerals Yearbook Area Reports 1956*, 3:364.

14

Other Dredge Grounds and Snake-Salmon Rivers

All along the Snake River and, to a lesser degree, along the Salmon, late-nineteenth and early-twentieth-century efforts to apply the new dredge technology met with little success. Fine gold consistently eluded recovery. With the passage of time, more sophisticated machinery was much more productive in the high placer regions—Pierce, Florence, Elk City–Newsome, Warren, Stanley Basin, Salmon City, Boise Basin, and the South Fork of the Boise River. Prichard Creek, at a much lower altitude, was also a fruitful area for gold dredging. In the more isolated regions, such equipment was sometimes applied with success and sometimes with none.

Apparently as early as the fall of 1894, a dredge of some sort was installed on the South Fork of the Payette River, just north of Boise Basin. A. E. Young of Buffalo, New York, purchased an interest in the Payette group of mines in the basin a year later, then bought the engine and boiler of the now-defunct gold dredge and moved them to the Idaho City area to run ore-crushing machinery.[1]

In the summer of 1899, several Chicagoans connected with dredging in Bannack, Montana, also made an effort to establish a dredge in the same

DOI: 10.5876/9781607324751.c014

general locale. At that time, in the aftermath of the Spanish-American War, the Boise *Idaho Statesman* reported that the local owners of the General Shafer, Admiral Dewey, and Homestake placer claims—Lee and D. F. Bunch, C. F. Dill, R. S. Baker, and several others—had bonded the property to Lawton C. Boney, Layman M. Paine, B. F. Graeter, and "R. Curren." Boney was a successful Chicago investment banker, Paine was a prominent lawyer, and both were officers of the Gold Dredging Company, the firm that built the *Fielding L. Graves*, the first successful bucket dredge in Montana. Graeter was no doubt related to A. F. Graeter, a Bannack merchant who had earlier invested in and given his name to the state's second dredge, the *August F. Graeter.* "R. Curren" was René Curin, who had earlier managed construction of the *Graves* and had "mysteriously disappeared" from Bannack in early 1899. The bonded ground was located on the South Fork of the Payette River and ran from two miles below the Garden Valley bridge to Bunch's ranch. Machinery for prospecting was brought in, and 10,000 feet of lumber were purchased from Hiatt's Placerville sawmill. Testing would begin immediately near the Bunch ranch, three miles above the bridge. Curin was managing the property and actively preparing to work it. He showed up in the Idaho setting, had put together a scow with an engine and pump to be used for prospecting, and had a crew busy widening roads and building several bridges between the placer ground and Placerville to expedite the transport of lumber and dredge machinery, which were expected to arrive in Boise soon.[2] Since nothing further was published about this endeavor, it is safe to assume that the property did not prove rich enough to warrant further development.

Quoting a Boise newspaper, the *Salt Lake Mining Review* reported in early 1902 that A. H. Butler and Archie Edmond had left overland for Payette Lake. Edmund owned a hotel at Lardo, a small town that was ultimately absorbed by McCall. Butler was in charge of building a large dredge for J. C. Blood and associates at Gold Fork, a tributary of the North Fork of the Payette River.[3] A dozen years later, in late July 1915, Mr. and Mrs. W. E. Pierce of Boise enjoyed an outing on Gold Fork Creek, eight miles from Donnely. It was reported that they both savored the vacation immensely. Mrs. Pierce was unwell when they left, but a long hike and the mountain air left her completely recovered. The excursion combined business and pleasure. Pierce had a joint interest in placer ground on the stream with the same J. C. Blood, now a more experienced mining man, and the two were trying to determine whether the 1,500 acres they controlled were suitable for dredging. Together, they had made the positive decision that dredging would begin the next year, with three machines set to work when the season opened.[4] No dredges subsequently appeared, but at least Mrs. Pierce had reclaimed her health.

To some investors the Secesh River, an upper tributary of the South Fork of the Salmon River, seemed to promise much. Relatively close to, but not part of, the Warren district, it, too, was something of a latecomer. In the spring of 1929, a Spokane newspaper noted that drill and shaft tests showed from twenty-two cents to twenty-eight cents in gold per cubic yard, with gravel twelve feet to forty-five feet deep on the ground of the Secesh Dredging, Mining, and Milling Company near Burgdorf, according to A. J. Draper, one of the directors. The firm had 3,680 acres along the Secesh River and its tributaries, Lake Creek and Grouse Creek, Draper explained, and fifteen men were continuing to prospect, as well as getting out timber for buildings and a plant. "We will put in a seven and a half foot close connected California bucket dredge, which will cost the company about $175,000," Draper said. Secesh Dredging was a Spokane corporation, headed by C. E. Inskip of Wallace as president. J. N. Rice and S. S. Bassett, both of Spokane, served as vice president and secretary-treasurer, respectively. Mrs. M. B. Morrow of Sumner, Washington, whose husband had located the property, was also listed as one of the directors. In a subsequent reincorporation under the laws of Idaho three years later, the headquarters were shifted to Wallace.[5] Whether there was any actual dredging in that interim period is not indicated. Certainly, the Depression years after October 1929 would have made capital difficult, if not impossible, to raise.

The US Bureau of Mines reported in 1946 that a new operator, the Secesh Dredging Company, had put a two-and-a-half-foot bucket dredge on Secesh Meadows in the same district, but it had dug for only one month. The boat worked longer the next year, running from May 15 to October 1, when it handled 50,000 cubic yards of gravel, for a yield of 277 ounces of gold and 49 ounces of silver. Since it was obviously not a paying proposition, the company did not attempt a third season.[6]

Another possibility was the Cariboo mining district in Idaho in the southeast corner of the state, in Bonneville County, forty miles north of Soda Springs. The paydirt was in several districts around the base of Cariboo Mountain, which reached an altitude of more that 9,800 feet. With both placer and deep-level quartz deposits, the area went through something of the usual pattern: a small rush, gradual working out of the richer surface ground, entry of the Chinese, and finally a limited application of twentieth-century technology. Progress was erratic, in part because of the frequent scarcity of water for sluices and hydraulicking.[7]

In 1921 five placers in one of the districts produced gold valued at $2,294 plus four ounces of silver. One of the contributors to that small output was the Wolverine Placer Mines Company's small dipper dredge, which had started up the year before on the Rasmason property. The firm was based in Salt Lake City and had been incorporated in May 1917, with nominal capital of $10,000 in 1 million shares of a penny each. What other access to capital it must have had is unrevealed, but the description of the company claimed an equipment outlay of about $70,000. That included the electric dipper, a 150 horsepower hydroelectric plant, eighteen miles of transmission line, and the McCoy Creek group of fourteen unpatented claims. H. R. Rasmason was listed as president of the concern, which would indicate that he owned the property or had options on it. Employing eight men, the machine suspended work in 1922, and there was talk of resuming digging in 1924–25, but that does not seem to have happened.[8]

Bonneville County had experienced a number of abortive efforts along the Snake in the very early days of experimental dredging. Apart from the Wolverine dipper, the only other attempt that comes to light is that of Feather Placers, which, with an unidentified dredge, in 1962 produced 131 ounces of gold and 3 ounces of silver in the Pine Creek district of the county.[9]

Owyhee County was a familiar name connected to both gold and silver, although the dredge came late to that county. The American Gold Dredging Corporation operated a dredge along Meadow Creek near Triangle, in the Steele district, beginning in 1928. Equipped with a two-foot electric Yuba flume dredge, the company dug for about six months that year. Narrow, steep gravel channels prevented the machine from running at full capacity, but the firm was able to ship a large amount of bullion to the assay office in Boise. Production of Owyhee County increased by $8,765 in 1929, primarily because of the new Triangle boat, but it dropped again the next year, when a water shortage curtailed work during the summer months. It was even worse in 1931, when the dredge operated only three months total because of a continuing water shortfall, although the company managed to send about $10,000 worth of gold to Boise.[10]

In 1935, as gold prices increased, Jordan Creek Placers installed a new three-foot portable Yuba on the stream of the same name in the Carson district. Starting up on July 25 not far from Silver City, the boat ran successfully until December, 1938, when its ground was worked out. In early 1937 it was joined on Jordan Creek by Frank Brendel's De Lamar Placer's dragline, which plowed through the village of Wagontown, leveling the buildings and leaving only the cemetery. In 1939 the Morrison-Knudsen Company, Ltd., a new operator in the district, began work with a power shovel and a dry-land washer.[11]

The Dixie district is located in the north-central part of the state, in Idaho County, inside the boundaries of Nez Perce National Forest and about twenty miles south of Elk City. It includes the upper basin of Crooked Creek, a stream that flows southwesterly into the Salmon River. That part of Crooked Creek would experience some successful dredging in the mid-1930s. Gold was first discovered in the area in the summer of 1861, no doubt by a Southern sympathizer during the Civil War, who promptly applied the name Dixie to the camp. For twenty years or so, no mining was carried out in the locality; not until 1884 was there even modest activity. Something of a revival occurred in the 1890s, which continued until about 1910. Then came a long stretch of inaction. Free gold existed in paying quantities only at shallow depths, and the district was so remote that transportation costs were prohibitive. As late as the 1930s, it was still eighty miles to the nearest railroad at Grangeville.[12]

Crooked Creek drains the entire district, which is within the Clearwater Mountains. The placer deposits were found at an altitude of 5,050 feet at the foot of Dixie Meadows to an altitude of 5,900 feet in Little Dixie Gulch. When S. R. Capps and Ralph J. Roberts did their survey in 1937, the pits and dumps of earlier efforts were still visible at several places between the town of Dixie and the Meadows. So was a more recent attempt to use a dragline scraper to cut a bedrock drain. Although a half-mile-long ditch was opened, so many boulders were encountered and the gold content was so low that the undertaking was abandoned without processing any gravel. Who was behind this particular work is not indicated.[13]

Dixie Meadows had possible dredge-able ground. Above the deep lower canyon of Crooked Creek, the Meadows offered a wide expanse of flat land a mile-and-a-half long. Parts had once been prospected, but modern equipment and the price of gold now made it tempting. Its location was near the

Dixie Ranger Station, however, and served both as a pasture for US Forest Service pack animals and as a landing field for the district, all of which meant the ground would likely never be open to dredging.[14]

In early August 1935 L. J. (Bill) Burrows announced that had obtained "one of the finest areas of placer ground in Idaho"—the Dixie placer on Crooked Creek. It was not for sale and no interest in it was for sale, he told the press, and the purchase of machinery and the expense of operation had been financed in the East. The ground consisted of 157 patented acres, which ran along the stream for four miles. The town site was part of the property but was exempt from the patent. Burrows had already built a summer home there. The average depth of the gravel was twelve feet, with the gold heaviest close to bedrock. A dragline of the new Northwest crawler type was on order, along with a washing plant built by Union Iron Works of Spokane.[15]

According to Capps and Roberts, the sixty-six horsepower dragline with a forty-five-foot boom began work in 1935, delivering gravel to the hopper of a separate washing unit that included a trommel and four sets of sluice boxes, lined with burlap and covered with wire netting. Tailings were carried away on a stacker belt. A photograph of the working dragline indicates extremely high pilings of the tailings because during the day shift both the dragline and the washing plant operated, but during the night shift they did not. Instead, the dragline aided a Caterpillar bulldozer in stripping off vegetation and barren material and in stacking the waste much higher that the stacker was able to do. Because of dry summers, water for the sluices was pumped from a sump and recirculated.[16]

With a seven-man crew (four on the day shift and three at night), in 1936 Dixie Placers, as the company was called, handled 190,000 cubic yards of gravel. It was expected that this "semi-dredge operation" would make its usual early start the next spring and do equally well, and the company did indeed maintain its gold production in 1937. By September 1, a cumulative total of about 400,000 yards had been washed, with an average value of around twenty cents a yard. One problem was the presence of many large granite boulders, which hampered progress. In one instance they were so bad that an 800-foot stretch had to be bypassed between Boulder and Dixie Creeks.[17]

By 1937 F. M. Beller had joined Bill Burrows in the ownership of Dixie Placers, or perhaps he was the easterner who had originally funded the operation. In any event, by that time the two men estimated that they had enough ground left to keep them busy through the summer of 1938 but that unless adjacent property proved worthwhile, the machinery would have to be moved elsewhere. For some reason, the company's dragline remained idle in 1938, but it was active during the next four seasons. In Dixie, a working season was relatively short: it opened between April 15 and May 1 and closed sometime in October each year.[18] The firm never reappeared after World War II.

From time to time there were other dredging efforts, none of which was successful. During the summer of 1937, Olive Creek Placers put what was described as "a 3/4-yard, gas-driven steam shovel" and a locally built washing unit on upper Olive Creek. Under the management of J. L. Stanley, a three-man crew stripped muck and old tailings from a cut about 200 feet long and 30 feet wide but went no farther. Because of poor machinery, they later abandoned the operation.[19]

In the same year, Idaho Placers did some dredging in Hundred Dollar Gulch. Using a ½-yard dragline scraper with a 35-foot boom and a standard washing plant, manager M. G. Read directed the digging of a cut that varied from 50 feet to 75 feet wide and from 8 feet to 10 feet deep, which ran upstream for more than 3,000 feet. The outfit did recover some gold, but in mid-August 1937, dredging was discontinued when the ground became unprofitable. Farther upstream, someone had also worked with a slack-line scraper but soon quit when it became apparent that there was little precious metal in the gravel. Also, when Capps and Roberts visited the Dixie district in 1937, preparations were under way by the Alpha Placer Company to rework old tailings and some new ground on Nugget Creek. The owners, C. M. Marts and A. R. McInroy, were building a dragline with a half-yard bucket, powered by an automobile engine, and were also constructing their own washing unit, which they hoped would be ready for operation in mid-September. *Minerals Yearbook* for 1938 did mention dragline recovery of gold on Alpha property.[20] In general, however, the Dixie district provided slim pickings for dredges. Only Dixie Placers managed to stay afloat.

Finally, in the far northern part of the state, after a long wait, dredging came to the Hoodoo mining district—an area of about twenty-eight square miles in northeast Latah County, on the headwaters of the Palouse River at the point where it leaves the Hoodoo Mountains. There is some uncertainty as to when gold was first discovered on the South Fork of the river, but it was probably sometime in the early 1860s. Whenever it was, the surface deposits were quickly exhausted, and, as in other placer regions in the state, white miners were replaced in the 1870s by the Chinese, who reworked the ground. A brief revival came in the placers in 1884, but it soon passed. The revival was followed by a decline, with little activity until 1893, when new strikes brought an upsurge in interest and an influx of new miners and promoters eager to join the ranks of the wealthy.[21]

In that year, for the first time in the district, entrepreneurs thought in terms of consolidating large blocs of placer ground along the river and working it with heavy machinery. Investors from Spokane and Minneapolis raised capital, while the group was represented locally by W. H. Muncy and J. N. Muncy from Tekoa, Washington. They located or took options on four miles on the North Fork and Palouse Rivers. Apparently, the plan was to operate a dredge on the stream. What kind of rig they had in mind is not known, but it would have had to have been something powerful enough to dig to bedrock at least eighteen feet down, a challenge in itself at a time when gold dredging was in its infancy. Another problem was the need to build a five-mile-long access road.[22] The Panic of 1893 and the depression that followed might have meant more men for the mines and cheaper labor, but it also created a scarcity of capital. In any event, the proposal did not mature, and nothing happened.

Again, in 1899 new interest was displayed from a different direction. At that time the Bucyrus Company of south Milwaukee, a well-known manufacturer of steam shovels and dredges, sent a representative to the Hoodoo district. The agent took options on considerable ground, which Bucyrus planned to prospect thoroughly the next year after the water in the Palouse River dropped to a low level. Apparently, the company intended to put a dredge in the ground if any of the property seemed rich enough to warrant it. Whether the testing actually occurred or whether corporate officials had second thoughts is unknown, but no action was taken in this case.[23] It would be a long wait—more than forty years—until modern mining technology arrived.

In 1940 there was finally a newcomer to the Hoodoo district. Five years earlier, the Engineers Gold Mining Company had been organized to locate and acquire dredge-able placer ground. The group subsequently purchased 280 acres of patented land and six placer claims, located above the town of Harvard, the nearest rail center. In the summer of 1939 they test-drilled the ground and then leased it to L. J. (Bill) Burrows of Spokane. Burrows formed Northwest Goldfields, Inc., of which he was president, as the operating company.[24]

During the winter of 1939–40, heavy dismantled machinery was laboriously trucked over a muddy dirt road along the Palouse River; the machinery frequently became bogged down in swampy ground. Frank Dilbert, a local miner who would serve as maintenance man on the dredge, was hired to pull the heavily loaded trucks out of bad spots with a Caterpillar tractor connected by a cable. Earlier Chinese placer mining had left a deep indentation at the site chosen for construction, which was easily adapted to form the pool. The dredge itself was a four-foot bucket-line machine powered by two 300 horsepower Cummings diesel engines. What company manufactured it and when is not known. Starting up on May 30, 1940, and digging until December 31, Northwest Goldfields worked on the North Fork of the Palouse River, beginning near the mouth of White Pine Creek. Virtually all of the placer gold and silver recovered in the district that season came from the new boat. The same was true in 1941, when the boat handled 1,021,726 cubic yards of gravel between January 1 and October 15, at which point the government shutdown halted production. The government also confiscated the two diesel engines for the war effort. Whereas the Hoodoo district's output for that short season had been 6,012 ounces of gold and 329 ounces of silver, practically all of it from the dredge, the take for 1943 was zero. Northwest Goldfields did not resume operations until 1947. Outfitted with a new 600 horsepower diesel engine, it started up again on the North Fork of the Palouse River, dug for three months, and took out 722 ounces of gold and 42 ounces of silver before exhausting its available ground.[25]

The dredge worked both virgin gravel and tailings. The average values were calculated at about 18 cents per cubic yard, while operating costs ran 7.224 cents a yard before depletion; depreciation and clearing the ground accounted for another 1 cent. Merle Wells of the Idaho State Historical Society estimates that the Northwest Goldfields boat produced slightly more that $550,000 from its five-mile voyage on the North Fork of the Palouse. Later, the company leased the dredge to Harold Behrens, who ran it intermittently for a short time in White Pine Gulch and on part of Doorman Creek before he and his brother switched to a dragline and a washing plant.[26]

The frantic efforts to apply the new dredging technology on the Snake and Salmon Rivers in the early years ultimately waned but did not die. Sporadic placer mining continued along both streams long after the black sand boom. So did discussion by geologists and mining men of the fine particles of gold and how to go about saving them, questions that were never satisfactorily answered. In addition, time did not lead to profitable means of handling the black sands. "New" methodology and technology were occasionally applied, but they never resolved the issues. Cyanide and flotation were not successful financially; the large investment in equipment meant high overhead charges. Everyone agreed that there was gold all along both rivers, but only a few were willing to continue the effort to capture it through large-scale efforts. The exceptions were usually inventors of new kinds of equipment.

The Idaho Dredging Company was organized in Spokane in early October 1928 to work placer ground somewhere on the upper Salmon River, according to President Charles LeMarinel, who also had plans for the Stanley Basin. The firm had valuable ground, LeMarinel said: "It is almost level and ideally situated for dredge operation."[27] He never identified the site, and nothing happened.

On the Salmon near the town of Lucile, a few miles north of Riggins, a good deal of placer activity occurred in 1929. One firm, the Seelridge Gold Mining Company, installed a dragline, along with a mill building and concentration tables, but apparently did little or no operating.[28]

In the summer of 1932, A. R. Brewer, a Spokane mining engineer, was convinced that the old bed of the lower Salmon in Nez Perce County was rich and "would return $1 a pan in gold" if the necessary equipment were installed—namely, a small dredge or dragline. With such machinery, which might cost as much as $5,000, Brewer believed a crew of five or six men could take out $600 a day, but further prospecting was needed to ascertain the extent of the deposit. Nearly five years later he announced plans to build a small suction dredge on the lower Salmon, presumably in the same area. If new explorations proved successful, he said, with the backing of local capital he would immediately construct a larger model for about $3,000, which would handle 240 cubic yards a day—figures that belied the Spokane newspaper headline "New Big Dredge to 'Rob' River." The machine would be mounted on a scow and would work the rich bottom of the river, on which test runs showed values of $2 to $8 per yard. "The ground increased in richness with depth and the gold increased in coarseness," Brewer explained.[29] His vision likely did not become reality.

Geologists Philip J. Shenon and John C. Reed were leaders of the joint National Geographic Society–US Geological Survey Salmon River Expedition in the fall of 1935 on what was billed as the "Wildest Boat Ride in America." Near the mouth of the Middle Fork of the Salmon, while camped near the mouth of Bear Basin Creek, they viewed "a few hundred feet below some inventor's dream of a short cut to riches, a suction dredge. Since the machinery was still in place, it appeared that the inventor had only lately been rudely awakened."[30]

At about the same time, two Spokane men, Albert Manning and Fred Schindler, leased Horseshoe Bend bar on the Salmon about eight miles above White Bird, then over the next two years they proceeded to install heavy mining equipment on the west bank of the river. Included were a permanent washing plant and a three-quarter-yard Erie steam shovel, the latter transported across the stream by a makeshift ferry. The function of the shovel was to strip away a deep overburden, as well as to load gravel into dumptrucks, which in turn conveyed it to the washing plant. By late April 1937 about 45,000 cubic yards had been processed.[31] How long the operation lasted is uncertain.

In 1938 several more attempts were made to dredge on the Salmon River. In the spring of that year, Frank Jones and associates shipped equipment to Large Bar in the Camp Howard district. Jones began with a donkey engine and a one-and-a-half-yard shovel dragline. His machinery was mounted on skids, and a large pump supplied water from the river. Tailings were pushed away by a bulldozer.[32] How Smith and associates fared was not made public. In August 1938 Roy Green and Max Kuney started up an operation on Slate Creek, a short tributary of the Salmon River. They used a gasoline-powered shovel, several Caterpillers, trucks, and a washing plant on ground where bedrock was from twelve feet to twenty feet deep.[33]

A new type of dredge was put forth in 1940 by its inventor, William Schierding, of Metaline Falls, Washington, who was also president of the Idaho Gold Dredging Company. According to Schierding, the new machine was able to penetrate to great depth, classify material under water, and bring up only fine gravel or black sand. Powered by gasoline or diesel, it could be built for a minimum price of $4,000, could be operated by one to three men, and would make its first run along the Salmon near the town of White Bird, about fifteen miles south of Grangeville, the company's headquarters. Prior to this effort, in the mid-1930s, with much publicity, he had applied his invention on the Pend Oreille River in his home state.[34] Both times, after the fanfare was over, so, apparently, was Schierding's fantasy.

The Snake River never again saw the kind of frenzy regarding dredging technology that had pervaded the early twentieth century. Occasional projects were announced, but, like those on the Salmon, none of them was successful. In the summer of 1927 the Idaho Power Company announced that its dredge in the Snake near Thousand Springs would begin operation as soon as the water level lowered enough to permit working the bars. The company had leased twenty acres on the south bank and estimated that it would take three years to complete the task.[35] This was probably a conventional dredge—suction or bucket line—but there was no description and no follow-up, which likely meant the machine was not able to recover the fine gold.

In the fall of 1934, geologists examined a number of placer operations of different kinds between King Hill and American Falls, along one of the most productive stretches of the river. One such project was located a few miles above "the Narrows," where the Reverend Ratschkowsky was dredging with a suction pump mounted on a float but with indifferent success.[36] In the early 1930s machinery was installed near King Hill to work the gravel with a steam shovel and separate the gold by flotation. The run was short-lived, and the plant was dismantled. In 1938, using a gas-powered shovel and a special screening and washing plant, the Northwest Mining Company treated 56,000 yards of gravel from the Rose Placer on the Snake River near King Hill.[37]

Of the Snake River placers, as late as 1945 William W. Staley, professor of mining engineering at the University of Idaho, wrote: "It appears that every conceivable kind of apparatus has been tried—bucket dredges, suction dredges, dragline scrapers, dragline shovels, picks, shovels, and wheelbarrows, etc. With the exception of one instance, there is no record of the economic success of any of the large-scale ventures." The one exception was the Sweetser-Burroughs suction dredge.[38]

To date, said Staley, based on meager information, small-scale operations requiring little capital outlay for equipment appeared to have been the most successful, with the bulk of known deposits averaging ten cents to thirty cents a yard. In 1945 no data were available on several of the dragline operations in the late 1930s and early 1940s, after the price of gold had been raised to thirty-five dollars an ounce, although it was known that the production of gold went up during that period.[39]

Staley pointed out that historically, "the greater part of the River has produced more or less gold," but development had been hampered by the fineness of the colors and an inability to save and process them economically on a large scale. Future success would depend on more careful sampling and determination of yardage, then the choice of proper mining and metallurgical techniques. Since extensive gold-bearing yardage was essential for profits, Staley preferred the suction dredge or the dragline excavator over the bucket-line dredge, primarily because of lower initial and operating costs. While not disparaging the burlap table, which claimed better than 90 percent recovery, he suggested a more modern and less bulky approach known as coarse-sand flotation, which promised to save close to 100 percent and was especially good for black sands and black sand concentrates rejected by other methods.[40] That suggestion never panned out, however.

Notes

1. *Idaho Daily Statesman* (Boise), May 30, 1895.
2. Ibid., July 13 and 23, 1899; *Dillon (MT) Examiner*, January 1 and 15, 1896; February 8, 1899; *Directory of Directors in the City of Chicago* (Chicago: Audit Company of New York, 1902), 23, 188; John W. Leonard, ed., *The Book of Chicagoans* (Chicago: A. N. Marquis, 1905), 72, 446. The *Statesman* misspelled the surnames of Graeter, Boney, Paine, and Cumin.
3. *Salt Lake Mining Review* 3 (February 1902): 22.
4. *Idaho Daily Statesman* (Boise), August 2, 1915.
5. *Spokesman-Review* (Spokane), June 24, 1929, clipping; *Spokane Chronicle*, September 23, 1932, clipping, both in Kimble Database, WSU.
6. *Minerals Yearbook, 1946* (US Geological Survey), 1441; *Minerals Yearbook, 1947* (US Geological Survey), 1409; *Minerals Yearbook, 1948* (US Geological Survey).
7. Merle W. Wells, *Gold Camps and Silver Cities*, Bulletin 22, 2nd ed. (Moscow: Idaho Department of Lands, Bureau of Mines and Geology, 1983), 99, 101–2; *Twenty-First Annual Report of the Mineral Industry of Idaho for the Year 1919*, 125.
8. *Mineral Resources, 1921* (US Geological Survey), 410; Stewart Campbell, *Twenty-Fourth Annual Report of the Mining Industry of Idaho for the Year 1922* (State Mine Inspector's Report), 59; Wells, *Gold Camps and Silver Cities*, 102.
9. *Minerals Yearbook Area Reports, 1962* (US Geological Survey), 3:355.
10. *Mineral Resources, 1928*, 673; *Mineral Resources, 1929* (US Bureau of Mines), 395; *Mineral Resources, 1930* (US Bureau of Mines), 643; *Mineral Resources, 1931* (US Bureau of Mines), 460; *Engineering and Mining Journal* 141 (May 1940): 29.
11. Yuba Manufacturing Company, *Yuba Dredges* (n.p.: 1939); *Engineering and Mining Journal* 141 (May 1940): 29; *Minerals Yearbook* (US Geological Survey), *1936*,

400; *Minerals Yearbook, 1937* (US Geological Survey), 325; *Minerals Yearbook, 1939* (US Geological Survey), 332.

12. "Mining in Idaho," Idaho State Historical Society Reference Series 9 (Boise: Idaho State Historical Society, 1985), 8; S. R. Capps and Ralph J. Roberts, "The Dixie Placer District, Idaho," Idaho Bureau of Mines and Geology, Pamphlet 48 (Moscow: University of Idaho, May 1939), 3–5.

13. Capps and Roberts, "Dixie Placer District," 5, 18.

14. Ibid., 18–19.

15. *Spokesman-Review* (Spokane), August 4, 1935, clipping, Kimble Database, WSU.

16. Capps and Roberts, "Dixie Placer District," 19, 19-A.

17. Ibid., 4, 19; *Minerals Yearbook, 1936*, 396–97; *Minerals Yearbook, 1937*, 323; *Lewiston (Idaho) Morning Tribune*, March 21, 1937, clipping, Kimble Database, WSU.

18. Capps and Roberts, "Dixie Placer District," 19, 21; *Minerals Yearbook, 1938* (US Geological Survey), 352; *Minerals Yearbook, 1939*, 328; *Minerals Yearbook Review of 1940* (US Geological Survey), 344; *Minerals Yearbook, 1941* (US Geological Survey), 350; *Minerals Yearbook, 1942* (US Geological Survey), 380; E. D. Gardner and Paul T. Allsman, "Power Shovel and Dragline Placer Mining," Bureau of Mines, Information Circular 7013 (May 1938), 42.

19. Capps and Roberts, "Dixie Placer District," 22.

20. Ibid., 22–23; *Minerals Yearbook, 1938*, 352.

21. A. H. Koschmann and M. H. Bergendals, "Latah County Idaho Gold Production," July 16, 2009, http://www.westernmininghistory.com/articles/112/page1; "Mining in Idaho," Idaho State Historical Society Reference Series 9, 17.

22. Richard C. Waldbauer, *Grubstaking the Palouse: Gold Mining in the Hoodoo Mountains of North Idaho, 1860–1950* (Pullman: Washington State University Press, 1986), 29.

23. *Mining: Journal of the Northwest Mining Association* (Spokane) 5 (June 1900): 3.

24. Charles R. Hubbard, "Mineral Resources of Latah County," Idaho Bureau of Mines and Geology, County Report 2 (March 1957), 10.

25. Waldbauer, *Grubstaking the Palouse*, 34–35; *Minerals Yearbook Review of 1940*, 345; *Minerals Yearbook, 1941*, 351; *Minerals Yearbook, 1942*, 381; *Minerals Yearbook, 1943* (US Geological Survey), 371; *Minerals Yearbook, 1947* (US Geological Survey), 1409.

26. Hubbard, "Mineral Resources of Latah County," 10–11; Waldbauer, *Grubstaking the Palouse*, 34–35: "Mining in Idaho," Idaho State Historical Society Reference Series 9, 17.

27. *Spokane Chronicle*, October 8, 1928, clipping, Kimble Database, WSU.

28. *Idaho County Free Press*, June 30, 1976.

29. *Spokane Chronicle*, October 8, 1928 (headline); August 26, 1933, clippings, both in Kimble Database, WSU.

30. Philip J. Shenon and John C. Reed, "Down Idaho's River of No Return," *National Geographic Magazine* 70 (July 1936): 94, 108.

31. *Spokesman-Review* (Spokane), April 25, 1937, clipping, Kimble Database, WSU; *Minerals Yearbook, 1938*, 322; *Minerals Yearbook, 1939*, 352; *Minerals Yearbook, 1940*.

32. *Spokesman-Review* (Spokane), March 5, 1938, clipping, Kimble Database, WSU.

33. *Idaho County Free Press*, June 30, 1976.

34. *Mining Journal*, March 15, 1940, clipping, Scrapbook, Yuba Manufacturing Company Papers, Merian Library, Chico, CA; *Metaline Falls (WA) News*, November 14, 1935; *Spokesman-Review* (Spokane), December 19, 1936, clippings, both in Kimble Database, WSU.

35. *Piute (UT) County News*, July 22, 1927.

36. Thomas H. Hite and Gerald Ashley Waring, "Gold Placer Mining on Snake River in Idaho," *Economic Geology* 30 (September 1935): 696–97.

37. Ibid.; *Minerals Yearbook, 1938*, 351–52.

38. Staley, "Fine Gold of Snake River," 7.

39. Ibid.

40. Ibid., 5, 7–8, 10.

15

The Struggle for Dredge Control

In general, for much of the twentieth century, gold dredging was acceptable to the public, in part because in numerous instances it revitalized declining mining districts by reworking ground left barren by earlier hand-mining methods or hydraulicking. True, those who lived near the big, ugly machines might not have cared for the unsightly dredge leavings or their constant raucous racket. As one poet put it:

> Oh! this the whine of the bucket line—the plait
> of grinding gear—
> The song that lifts through graveyard shifts by
> hour, day, and year—
> The anguished grind as buckets bind in stress and
> strain and shear.[1]

This was part of the price of progress, jobs, resource development, and prosperity. Concern for the environment was not yet popular.

DOI: 10.5876/9781607324751.c015

By the early 1930s this attitude had begun to change, and in Idaho, as in placer mining areas elsewhere in the West, opposition voices began to call for state legislation to halt the pollution of streams and despoiling of the land by dredging. The increase in the price of gold in 1935 brought more bucket dredges and draglines, with high production until World War II, which made the problems more visible, as did the Cold War programs subsidizing dredging for critical minerals.

The movement for dredge control was long and gradual, and it eventually embraced a number of groups. The farm element was never ambivalent about dredges and the environment. In the spring of 1938 W. H. Tuller, manager of the Boise River Board of Control, noted that reclamation leaders in southwest Idaho strongly opposed a proposal to install a gold dredge in the Boise River above Diversion dam. Officials of irrigation districts expressed the fear that the dredging operation would stir up silt that would be carried downstream into irrigation canals.[2] More than three years later, farmers along the river vigorously protested the application of the Nampa Gold Dredging Company to lease ground between the dam and the mouth of Mores Creek.[3]

Along with the farm element, led by the Grange, other segments of the economy joined the movement. In 1936 fish and wildlife groups in the state united as the Idaho Wildlife Federation, committed to work to limit the impact of mining on fish and the outdoor recreation industry. The Idaho Outdoor Association also took an active role, along with conservationists, the press, and, ultimately, politicians and private citizens. The battle was an uphill effort, with the mining industry usually in command. It took time for the various groups to coalesce, and the struggle was fraught with obstacles and a number of failed legislative efforts. The Idaho Mining Association was a formidable foe.[4]

The first effort, which came in 1939, was a Grange-sponsored stream pollution bill in the legislature that contained the phrase "regulation of dredging of gold in Idaho rivers." The measure was designed to control many forms of stream contamination, including not only dredging and dumping other mining debris but disposing of municipal sewage as well. The bill was defeated by a large margin, in part because it was too ambitious; small towns and cities could not afford to meet the criteria required therein for sewage and water systems.[5]

In 1940 and 1941 the State Land Board approved two new dredges for the Boise River system, one for the South Fork and one for the Middle Fork. This brought immediate resistance from the growing dredge-control faction, especially sportsmen. One newspaperman pointed to the most polluted stream—Mores Creek, a tributary of the Boise River. He insisted that good fishing on the creek was long gone; dredging had turned it into "a soupy mass of muck year round . . . not a fish lives there." When the board sanctioned the second lease, the friction increased. The president of the Ada County Game and Fish League scoffed that the dredge already operating there had ruined the river for sportsmen, "that rocks are covered with slick silt half an inch thick; that fish will hold their noses and wiggle to freer waters." Those two complaints, plus one beginning to develop against the Yankee Fork dredge on the confluence of the Salmon, virtually ceased with the ban on dredging during World War II.[6]

In December 1946 Boise mining engineer Leverett Davis petitioned the State Land Board to grant him a ten-year lease for dredging the Boise River about ten miles downstream from Arrowrock dam. That request was put on hold pending a brief study of how such an operation might affect irrigation projects that drew water from the stream, but within a month the request was granted. In early 1947 Davis was given a ten-year lease on the river to operate a suction dredge, and he agreed to post a $100,000 performance bond to guarantee irrigation water users that sand and silt would not be present in the canals. One source pointed out that Davis's "precedent shattering ten-year lease" was approved by the Democratic State Land Board, which consisted of the governor, the attorney general, the secretary of state, the auditor, and the superintendent of public instruction. The conditions of the agreement cited by Spokane newspapers made no mention of a performance bond but included other stipulations. Davis agreed to pay the state 25 cents per acre along a ten-mile stretch between Diversion dam and Arrowrock dam, plus 2.5 percent of the gross proceeds of the dredging operation. Davis expected to start drilling within thirty days to determine whether the ground was worth his acquiring a large dredge that might cost as much as $2 million. The board had acted just four days before all of the members' terms expired, and the new Republican board that swept into office responded to farmers' complaints and chose to challenge the legality of the agreement.[7] The final details are blurred, but apparently the lease was nullified, and no announcement of a dredge active in that area was forthcoming in subsequent *Minerals Yearbooks*.

At its annual meeting in 1946, the Idaho Mining Association took the initiative when it announced that water pollution was "a local problem" to be handled by the industry itself, "without any compulsory legislation." The industry's solution was to bypass the main body of critics by creating the Water Utilization Committee as an autonomous body that ultimately failed.

In 1949 the Idaho House of Representatives bill sponsored by the Idaho Outdoor Association passed by a substantial margin but failed in the Idaho Senate when the mining committee tabled it indefinitely. After the 1950 Defense Production Act (DPA) had the effect of increasing the number of dredges in the 1950s, public reaction grew as well. Since the DPA emphasized minerals deemed crucial during the Cold War, the mining industry stressed the role of duty, patriotism, and fear. Harry Marsh, secretary of the Idaho Mining Association, declared that advocates of tough regulation were "ignorant of how important this dredging is to American defense; apparently they think they can stop the Russians by hitting them in the face with a fish." Another bill for dredge control, introduced into the legislature in 1951, suffered the same fate as the previous one. At the same time, cases concerning leases under DPA auspices began to attract sharp criticism. One such case involved Kate and Claude Finch, who applied for renewal of a lease on eighty-nine acres of streambed of the Salmon River for dredging monazite. Local citizens protested loudly, and residents of the Payette Valley took the cue and condemned another, similar lease on the South Fork of the Payette River near Crouch. Farmers along both rivers added their voices, and the press also joined the chorus. The State Land Board denied the Finches their lease. Idaho was poised for its first law to regulate dredge mining.[8]

When the statute was finally passed in 1953, it was a watered-down version, engineered by the dredge supporters in the legislature. It applied only to large machine operations able to handle more than a thousand cubic yards per day. The statute did provide for settling ponds, but only when "reasonably possible." Watercourses must remain unobstructed, and tailings had to be leveled. What the statute did not require was that streams be returned to their natural course or be rehabilitated by adding topsoil or plant life. Administration was in the hands of the state inspector of mines, with many details left to his discretion.[9]

The statute was a beginning, but the public demanded more. In 1954 the Idaho Wildlife Federation took the lead in seeking to bypass the legislature and going directly to the citizens with a ballot initiative to strengthen the new law. The initiative was successful by a majority of more than 85 percent of the vote and led to significant additions to the 1953 statute. It reduced the amount of gravel a dredge had to be capable of moving in a day from 1,000 to 500 cubic yards; thus, it applied to more machines. The initiative also specified that settling ponds be large enough to reasonably clarify the water emptied back into the stream, and it required that watercourses be returned to a fair facsimile of the original, complete with pools and natural meanders. Moreover, a dredge operator was now called upon to post a bond of $300 for each acre of land to be mined to ensure restoration.[10]

The amended law was tested almost immediately when the Porter Brothers Corporation installed its six-foot Yuba on Bear Valley Creek, a tributary of the Middle Fork of the Salmon River, in 1955. Bear Valley was located in high mountain country about thirty miles northwest of Stanley, an area ignored by all but a few early prospectors but well-known to fishermen in the mid-twentieth century. The Porters had respected the new regulations, in the belief that they represented the way to prevent even more drastic restrictions. They had built settling ponds, but when the Idaho Fish and Game Department analyzed the water below the dredge, it found it contained an average of 1,200 parts per million of suspended solids—five times the content allowed by California's dredge-control law. With some negotiations, Porter Brothers agreed to abide by the standards of the State Land Board. The monazite boats at Long Valley near Cascade followed the same course and for the same reason—leveling tailings, using settling ponds, and doing some re-vegetation. The fact that several other operators were not living up to the letter of the law strengthened the control exponents' desire for more stringent regulations A legislative amendment in 1957 failed to define water standards, but it did set a minimum bond requirement of $6,000 for each twenty acres and allowed for forfeiture of the entire amount if terms of the law were not met. Additional punishment was added in a clause that made noncompliance a legal misdemeanor.[11]

In January 1956 the State Land Board charged Vernon B. Finch and the Clearwater Dredging Company with breaking the 1954 dredge-control law while digging on the Crooked River in Idaho County. Eight moths later the board stripped Finch of his permit. Finch turned to the courts for resolution, and in September 1957 the Idaho Supreme Court handed down the final decision. It found that one part of the Dredge Mining Control Act violated the state constitution—the provision that allowed any appeal of a board order to circumvent the lower courts and go directly to the supreme court. Inasmuch as the ruling made it clear that the board's power to revoke a permit was "null and void," the dredge mining law was ineffective.[12]

The flaw in the 1954 law was not rectified immediately. Little gold dredging was being done by this time, and advocates had more important issues to address, especially opposing new dam construction and supporting a bill to create a wilderness area. By the late 1960s the federal government had either dropped or completed the dams disputed by the Idaho Wilderness Federation and other such groups. The national Wild and Scenic Rivers bill was passed by the US Congress in 1968 without any limitations on mining, so conservationists went back to work on their own control legislation.[13]

In the spring of 1968 a Texas outfit asked the State Land Board for a permit to dredge for gold and garnets in and along the upper St. Joe River. This was a small venture that contemplated moving only about fifty cubic yards of material a day, well below the capacity for regulation by the state's dredge-control law. Recognizing that interest groups were still fidgety where placer mining was concerned and despite the company's objections, the board held public hearings on the permit. After finding that a wide range of organizations were opposed to even a small operation in such a pristine setting, the opposition introduced a bill in the sitting special session of the legislature, which would require any rig capable of moving ten yards of gravel per day to have a permit. The proposed bill cleared up the board's right to hold public hearings and brought the appeals section into agreement with the state constitution. The amendment passed the house by a vote of fifty-nine to one but lost when the senate added an amendment that would have weakened the power of the State Land Office, and the special session ran out of time. Finally, when it became apparent that control advocates had the upper hand, in early 1969 the Idaho Mining Association, sensing defeat, submitted a bill containing a number of concessions. With this bill as a point of departure, the association and the Idaho Wildlife Federation worked through legislative committees to make it palatable to both groups. It became law in March 1969 and might be regarded as a compromise measure necessary to save what was left of the dredging industry. Another dredge bill was passed in spring of the following year that did away with dredge mining on Idaho rivers set apart as "wild and scenic."[14]

Later, these laws were challenged in the courts. Three Montanans—James Click Sr., Eugene Weiss, and Oral Lake—were dredging on unpatented public land on Sherlock Creek in St. Joe National Forest sometime in the early 1970s. In late 1973 the State Land Board was able to block their operation until they complied with the state's dredge-control statutes. The dredgers argued that federal mining laws took precedence over those of the state, but the Idaho Supreme Court decided otherwise and found the regulatory acts "valid and fully applicable."[15]

This left the state with regulations that had teeth, but in a sense they came too late. The bulk of the gold dredging was over, and the environmental damage had been done. The dredge-control laws added measurably to the expense of any possible large-scale operation. Only occasionally thereafter was any such venture attempted.

Notes

1. W. H. Gardner, "Song of the Gold Dredge," *Engineering and Mining Journal* 108 (August 16, 1919): 286.

2. *Spokesman-Review* (Spokane), May 8, 1938, clipping, Kimble Database, WSU.

3. *Engineering and Mining Journal* 142 (October 1941): 85.

4. Darin W. Vrem, "A History of Idaho's Dredge Mining Industry and the Dredge Mining Control Movement 1896–1970," MA thesis, Boise State University, Boise, ID, November 2004, 21, 28, 33. This is an excellent study of the subject in great detail.

5. Ibid., 25–26.

6. *Idaho Daily Statesman* (Boise), November 2, 1940 (quotes); *Boise Capitol News*, December 5, 1941, both quoted in ibid.

7. *Spokesman-Review* (Spokane), January 3, 1947; *Spokane Daily Chronicle*, December 6, 1946; January 28, 1947.

8. Vrem, "History of Idaho's Dredge Mining Industry," 28–31, 35.

9. *General Laws of the State of Idaho: Passed at the Thirty-Second Session of the State Legislature, 1953* (Boise: Syms-York, 1953), 291–94.

10. *Idaho Session Laws*, 1955, in Vrem, "History of Idaho's Dredge Mining Industry," 29–35.

11. Ibid.; Vrem, "History of Idaho's Dredge Mining Industry," 41–43.

12. *"State of Idaho v. Vernon B. Finch," Pacific Reporter* 315, 2nd series (September 10, 1957): 527, 529–30.

13. Vrem, "History of Idaho's Dredge Mining Industry," 43–45.

14. *Idaho Session Laws*, 1969, in ibid., 46–47, 49–50.

15. *"Cecil D. Andrus, Governor, et al., v. James Click Sr. et al.," Pacific Reporter* 554, 2nd series (September 24, 1976): 969, 972–73, 982.

16

Aftermath

The Cleanup

It was clear from the beginning that gold dredging wreaked havoc with the streams. Nowhere was this more obvious than in the upper part of the South Fork of the Clearwater River, where four major tributaries—the Red River, American River, Crooked River, and Newsome Creek—had experienced considerable dredging, as had the main-stem South Fork itself. Newsome Creek and Crooked River had the heaviest digging. The result in all these areas was radically altered channel courses, destruction of riparian vegetation, and elimination of proper fish habitat conditions. It took a new generation with a concern for the environment to undertake rehabilitation of the damaged creeks and rivers. Since most of the streams in the Idaho dredging areas originally supported large numbers and diverse populations of fish species, including Chinook salmon, steelhead trout, bull trout, and cutthroat trout, when rehabilitation occurred the focus was usually on fishing and the fish habitat.

Gold dredging was not the only cause of the impairment of waterways. Natural phenomena such as floods and forest fires, as well as timber

DOI: 10.5876/9781607324751.c016

harvest, building and usage of dirt roads, and grazing, can all contribute to the problem.

The Clearwater River originates in the Bitterroot Mountains on the Idaho-Montana border and flows westerly across the state to Lewiston, where it joins the Snake. The eastern half of its drainage is primarily forestland, while the western portion is mainly private land and corporate timber ground. About 13 percent of the South Fork drainage is taken up by the Nez Perce Tribal Reservation. Sixty-three miles of the main-stream Clearwater are within the reservation, as are eleven miles of the South Fork. The entire drainage is part of the Native Americans' ceded lands. Thus, it is understandable that the Nez Perce Tribe played a major role in the effort to restore streams that had suffered damage to their fishing habitats as a result of gold dredging.[1] Since the Nez Perce National Forest overlapped tribal territory, a common interest readily created a partnership between the two entities in reconstruction work.

Restoration basically sought to return the stream to its native condition as much as possible after the flow had been modified in several ways. Often, dredging eliminated meandering and in the process cut new channels wider and shallower than the originals, which offered few deep holes and little cover for fish. After the trees, shrubbery, and overburden were removed, the dredge was relentless: it gobbled up vegetation, sand, gravel, and whatever else might be in its path. As hydrologists noted, the water in a straighter, faster-flowing stream, deprived of its shade, had been "thermal-modified" and was no longer suitable habitat for young salmon or trout. As the temperature in the stream rose, the amount of dissolved oxygen in it decreased, with a negative impact on the fish. As one hydrologist put it, "A meandering, bumpy creek is a cool creek."[2] Add to that the oil, silt, debris, and other contaminants a dredge brought to a river or creek. Then there were the tailings: all dredge tailings were not created equal. It depended upon what type of dredge or washing plant was used. A floating bucket dredge produced a stereotyped arc pattern, one arc piled against another, because the boat pivoted back and forth on its anchor—its spud. This type of machine usually left its trademark in the form of a tailing pool when the operation was complete. A dragline dredge stripped the overburden, washed the paydirt, then moved on to the next cut. When working with either a stationary or a floating washing plant, its residue was usually large mounds with a good deal of silt and vegetation on them, but sometimes, depending on the ground, the trail might be difficult to ascertain. If a power shovel with a stationary washer was used, the tailings might be saved to refill the pit or be dumped by truck near the processing plant. Thus, if the operation required frequent moves, from the air its leavings might have resembled a string of pearls.[3]

The bucket-line dredge's caterpillar-shaped tailings were the hardest on the environment because they turned the ground upside down, with the cobbles on top of the soil. Sometimes piled as much as fifteen feet deep alongside or in the stream itself, they formed a kind of armor on which vegetation would not grow, thus effectively eliminating any semblance of a floodplain. A single dredge boat might leave several hundred acres of such tailings in its wake.

The dredged-over areas of the Crooked River and American River on the South Fork of the Clearwater eventually came under close scrutiny for water quality with respect to fishing habitats. Under several different owners, the Mount Vernon dredge had dug on tributaries or on Crooked River itself and at other times had also run on the nearby Red River and Buffalo Gulch. When the time came, the Crooked River was a logical candidate for improvement of its fisheries habitat. Feeble efforts were made in that direction as early as 1962, when the Idaho State Highway Department asked for a permit to use Crooked River dredge tailings to surface local roads. A decade later, another 30,000 tons of tailings were taken out near Orogrande for road work, and a little vegetation seeding occurred at this time. In 1973 the US Forest Service set forth a challenging plan that would allow the highway department to remove an additional 10,000 tons; at the same time, substantial improvement of the stream would be done. The roads received their gravel, but only a meager amount of work was done to improve the river at this point.[4]

In 1980 the US Congress came to the rescue with the Pacific Northwest Electric Power Planning and Conservation Act, one part of which mandated that the Bonneville Power Administration (BPA) spend part of its profits to mitigate the adverse effects of the Columbia hydroelectric dam system on fish and wildlife in Idaho, Montana, Oregon, and Washington. To facilitate this, the law created the Pacific Northwest Electric Power and Conservation Planning Council, with members drawn from the four states. Since many of the Idaho streams affected by gold dredging fell within the Columbia Basin, they became eligible for financial aid from Bonneville Power for rehabilitation work.[5] Five goals included improved habitat for anadromous fish, improved habitat for resident fish, improved riparian habitat for wildlife, improved visual quality of the dredged area, and a proven example of stream rehabilitation as a pattern for other dissipated watercourses.[6]

Originally, two contracts with Bonneville covered projects in the Nez Perce National Forest. In 1984 these projects were combined into one, which outlined a seven-year schedule for restoration work in the South Fork Clearwater Sub-Basin, with a grant of $174,926. In the next seven-and-a-half years this agreement was amended eight times, adding project activity, extending the time period, and increasing the funding to a total of $1,181,730. Thus, in 1984, with a Bonneville contract and help from the Shoshone-Bannock Tribes and the Idaho County Soil and Water Conservation District, fisheries biologists from Nez Perce National Forest took the lead, with a focus on seven separate locations: Crooked River, Red River, Meadow Creek—all three of which had dredged ground—plus Haysfork Glory Hole, Cal-Idaho Glory Hole, Fisher Placer, and Leggett Placer.[7] In placer mining, the term *Glory Hole* usually meant a large hollow washed in the ground by a hydraulic giant nozzle.

Testing the waters was an early part of the project. US Forest Service tests found that total and dissolved water column concentrations of argon, copper, zinc, silver, and mercury were all below detection levels and that in dredge pond-bottom sediments, all of these elements plus iron, lead, and cadmium were also at safe concentrations for fish. The biologists proposed to realign the existing stream course by connecting six dredge ponds to the Crooked River, constructing new channels, and modifying an additional intermittent channel to assure perennial flows. The object: to provide "additional off channel rearing for the juvenile anadromous salmonid that inhabit the stream system"—meaning salmon that live in the sea but go up freshwater streams to breed.[8]

What followed was a complete rebuilding of sections of both the Crooked and Red Rivers. Between 1985 and 1989, more than 10,700 square yards of new floodplains were created by moving tailings away from the Crooked River where they constrained the channel. In most instances, erosion control matting was used to help collect silt. Through pond connections and installation of more than 789 cover-producing in-stream structures—that is, log or rock weirs, random boulders, deflectors of some sort, and large pieces of woody debris, which had to be anchored in some fashion—a total of 17,900 square yards of juvenile rearing and winter habitat was created. At least 30,000 hardwood shrubs and small conifers were planted in riparian areas along the river.[9]

But the work was by no means complete. As late as December 2012, the Nez Perce and Clearwater National Forests gave notice in the *Federal Register* of their intent to prepare an environmental impact statement for the Crooked River Valley Rehabilitation Project. Two miles of the valley had been heavily dredged earlier, leaving behind large tailing piles and deep ponds. Restoration would reconstruct the river channel to its original meandering course, as well as reroute a local road. On August 13, 2015, the BPA decided to fund this project—so it goes on.[10]

The Red River had also been dredge-mined, but less than the Crooked River. In sections of the stream, the riparian vegetation along the watercourse had been wiped out; with the lapse of time, it had only sparsely regrown. On parts of the Red River, too, dredging had obliterated the channel bed and banks. Here, restoration treatment followed the same pattern as that on the Crooked River. Shoreline was stabilized with the creation of new floodplains, 1,692 yards of side channels were added, and more than 503 yards of the river channel were restored and realigned. In addition, 319 in-stream structures were supplied to enhance the habitat, and about 11,000 trees and shrubs were planted. It was estimated that roughly 4.3 miles of the Red River would benefit.[11]

The work was still not finished, however. The project continued with Bonneville contracts to both the Nez Perce Tribe and the US Forest Service, with a total expense budget of $118,899 for 2004, $374,409 for 2005, $231,686 for 2006, and $324,727 for 2007. Records indicate that the Nez Perce Tribe contract started in 2002 was due to end in November 2013, while that of the Forest Service began in March 2007 and closed in November 2010. Apparently, work was done at several reaches on the Red River, one of which was a two-mile stretch upstream from the bridge below Dawson Creek. A private firm, Watershed Consulting, LLC, of Whitefish, Montana, gave a preliminary cost estimate of $601,529 for that part of the restoration.[12]

The restoration of the lower Red River Meadow seems to have had a separate life. The meadow was composed of four separate parcels of land—a total of 1,300 acres along about 4.4 miles of the stream. One of the four parcels came up for sale in 1993 and was purchased by a combined group that included the BPA, the Idaho Fish and Game Department, the Rocky Mountain Foundation, Trout Unlimited, the National Fish and Wildlife Foundation, and the US Forest Service. In 1994 the property was deeded to the Idaho Fish and Game Department, in an agreement with the BPA, to manage in perpetuity as the Red River Wildlife Management Area. The BPA then funded restoration, which went on in stages in the traditional manner for ten years. Apparently, only a little dredging had been done on a portion of the land. No mention of dredge tailings appears in the numerous descriptions of the renovation.[13]

Of all the streams flowing into the South Fork of the Clearwater River, Newsome Creek experienced the greatest intensity of dredge mining. It was dredged from the mouth of Nugget Creek almost to the mouth of Radcliff Creek—about half its length. Located a dozen or so miles northeast of Elk City, Newsome Creek flowed south for 15.7 miles before joining the South Fork. Since it was entirely within the ceded territory of the Nez Perce, the tribe was heavily involved in partnership with the Nez Perce National Forest in efforts to rehabilitate the stream.[14]

Over the years the tribe was also a party in a number of other damaged watercourses in central Idaho and Oregon, but it commenced simple stream restoration projects in Newsome Creek as early as 1997. The US Forest Service had taken a few steps in that direction in 1988, when it began stream monitoring and typing; the organization also placed a few in-stream structures in the creek a year later and re-vegetated an old placer mine in 1996. In these early efforts the Nez Perce had been funded by the BPA in a cost-sharing arrangement, a project "directed toward reclaiming raw soil and [stabilizing] the effects of historic placer mining." Of a total budget of $79,630, the US Forest Service contributed $45,981, while the remainder came from Bonneville through the tribe. In 2001, in conjunction with the US Department of Agriculture and the US Forest Service, the Nez Perce Tribe began a more sophisticated program of rebuilding fractured fishing habitats.[15]

The proposed Newsome Creek project consisted of two major components: restoration of a 4.1-mile section of the stream and the improvement and decommissioning of roads in the northeast part of the Newsome Creek watershed. The latter was to help alleviate sediment input. The state of Idaho had placed the stream on the impaired list in terms of water quality under the Clean Water Act. Dredging on the upper creek had left an estimated

85,000 cubic yards of tailings.[16] Damage to the stream was similar to that left by dredging elsewhere, and the remedies were the same.

As was the case in other waterway renovations, much of the financial support came from the BPA, which received a budget request in 2000 for $364,725 to fund a five-year cleanup of Newsome Creek. Nez Perce National Forest promised to provide $80,000 in technical expertise.[17] In March 2001 the tribe was awarded $1,081,324 for the project. Beginning in 2005, with a new agreement and a new Newsome Creek budget, Bonneville made its funding contingent on a 30 percent cost-share from the US Forest Service. BPA financing continued at varying levels for a number of years, and in 2011 it provided the Nez Perce Tribe with an additional $1,327,539, to be used through 2014 if needed.[18] Thus, after more than a dozen years, the tribe is succeeding in its drive to clean up Newsom Creek.

Another of the dredged-over streams that shared a somewhat belated rehabilitation was the Yankee Fork, one of the major tributaries of the upper Salmon River in the central part of the state. For more than six miles downstream from Jordan Creek the Yankee Fork channel had been seriously altered, with much of its natural meandering pattern eliminated, its riparian vegetation covered by tailings, several of its confluents isolated, and a number of pools disconnected. Moreover, mercury used to amalgamate the gold was a potential hazard, although any lost quicksilver would probably have been in the soil underneath the cobbles of the tailings. In addition, the lower 1.3 miles of Jordan Creek had also been dredged, probably with a dragline machine. What for centuries had been a choice habitat for Chinook salmon and steelheads was no longer such.[19]

The Shoshone-Bannock Tribes had made a few improvements on Yankee Fork beginning in 1984, under a separate, more general contract with the BPA—the Salmon River Habitat Enhancement. The tribes also did some restoration work there in 1987 and 1988, in a collaborative effort among the BPA, the US Forest Service, the Idaho Fish and Game Department, and Jack Simplot, who still owned the property that had been dredged. At this time the improvements were in the form of developing new channels to connect four ponds with the main stream, along with adding large rock, root masses, and some vegetation to provide cover for young fish. In 2001 the US Forest Service conducted a fish habitat inventory, which showed a lack of pools and high width-to-depth ratios. Spawning and rearing habitat were identified as in poor condition in the dredged area but in fair to good condition upstream of the dredge tailings.[20]

The Bannock Tribe had long been familiar with Yankee Fork. For many generations the tribe camped at the mouth of Ramey Creek, a tributary, to harvest spawning Chinook salmon.[21] Combined with the Shoshones, they would play a highly important role in the reclamation of Yankee Fork, as well as in reclamation elsewhere in Idaho. According to the Challis press, in 2012 the two tribes had a total of $61 million in contracts with the BPA for rehabilitation projects, mainly in the Salmon River drainage system.[22]

But the tribes were slow to convince the Independent Scientific Review Panel (ISRP), which screened potential BPA projects, that Yankee Fork could be remade into a viable fish habitat. In 2002 the tribal proposal was rejected, but two years later the BPA provided some funding for field data as a start. From 2004 through 2007, with cooperation from other state and federal agencies, including the Stream Ecology Center at Idaho State University in Pocatello, the tribes monitored the stream to collect water-quality information—pH level, dissolved oxygen, turbidity, temperature, and discharge measurements. A no-cost extension was awarded to finish the work covered in the 2007 contract, but that, too, was prolonged until the end of 2008. No funding contract was awarded for 2008–9 until the project provided an implementation plan and justified financial support. When the 2008 plan was submitted, the ISRP found it wanting, but after subsequent changes the panel agreed to divide the valley into three separate sections and allow the tribes to continue to work on one of them. If that proved successful, funding of the entire program would be considered. This was part of the Pacific Northwest Electric Power and Conservation Planning Council's Three-Step Review Process, designed to expedite very expensive projects. The tribes had estimated that future costs for 2010 would be $5 million and for 2011, $6 million to complete restoration of the Yankee Fork area.[23]

Meanwhile, in 2008–9 the so-called action groups—Bonneville Power, the US Army Corps of Engineers, and the US Bureau of Reclamation—signed agreements with a number of confederated tribes in the Pacific Northwest. The action groups also signed agreements with the Shoshone-Bannock Tribes and with the states of Idaho, Montana, and Washington. These agreements were known as the Columbia Basin Fish Accords. On June 30, 2008, the council received from the Shoshone-Bannock Tribes the Yankee Fork Flood

Restoration Project Implementation Plan for 2008–18, intended to address the three-step review conditions in the previous funding recommendation. That plan, too, was rejected by ISRP reviewers. Three months later the BPA on its own entered into a memorandum of agreement with the two tribes that included this project. In early 2012 the BPA sent word to the council that after working with the US Bureau of Reclamation, the US Forest Service, Trout Unlimited, and Jack Simplot, the tribes had restructured the proposal, addressed ISRP concerns, and provided for a longer time span. The ISRP subsequently agreed, as did the council. The accord budget for the project was set at roughly $5,177,259.[24]

Records of the Bonneville Power Administration's Division of Fish and Wildlife indicate that $160,373 was allocated to the tribe's Yankee Fork project in each of the years from 2004 through 2007, but nothing was paid in 2008. For 2009 the budget was stepped up to $313,042; for 2010 it dropped to $273,944, then was raised to $317,134 in 2011. The year 2012 saw a big boost to $1,770,200, then the amount fell back to $222,003 for 2013. The Idaho Office of Species Conservation had also been awarded $1,470,800 for the same project. The work was clearly still going ahead, but it was also evident that there was concern about the slow rate of progress. When the Northwest Power and Conservation Council met in Spokane in August 2012, it passed a resolution making further project implementation contingent on a favorable ISRP review of a revised project proposal. A month later the BPA announced that lower revenues from power sales might impact the fish and wildlife programs, which in the previous fiscal year had expenditures of $650 million for the four-state region.[25] The final fate of Yankee Fork restoration is yet to be resolved.

The Boise Basin probably suffered the most dredge damage to its streams. It had a long history of placer mining using primitive hand methods mingled with hydraulic monitors—powerful water canons—that directed water under great pressure to wash away the cliffs and hillsides into long sluices with relentless force. With debris from both the point of contact and after the gold had been removed, this was undoubtedly the most destructive kind of mining. Wherever it was used, it left a layer of tailings along or in the streams in the valleys and flatlands and often large, deep holes as well. When the bucket dredgers arrived, they often found themselves working one layer of tailings and depositing another on top.

As of July 1949, the Grimes Creek area between the ghost town Old Centerville and the town of New Centerville was the topic of at least some restoration. The ground had been owned and dredged over prior to World War II by the Baumhoff-Marshall Company, from which the Idaho Fish and Game Department had obtained a lease that included a stipulation that the public have free access to the improved area. The property had not been leveled and had been left with a mass of heavy tailings and limited vegetation growth, along with several ponds averaging about eight feet deep. The stream was subject to a sharp drop in volume and a rise in water temperature during the hot, dry summers. Its bed was shallow, with few deep holes, and afforded little cover for trout, although there were plenty of aquatic insects for a sizable fish population. Fishing during the early weeks of the 1949 season was only moderate, and during the remainder of the season fishermen were abandoning the creek.

Donald Andriano was hired in the summer of that year, probably by the Idaho Fish and Game Department, to make improvements in the situation. Andriano had a masters degree from Utah State University and went on to become chief of fisheries for the Utah Division of Wildlife Resources, as well as president of the Western Division of the American Fisheries Society, a national organization devoted to the betterment of fishing and fisheries.[26]

Improvements covered 2.3 miles on Grimes Creek. Work was done to increase both the surface area of the stream and the shelter for fish. To this end, the creek was diverted in and out of the various dredge ponds by building dykes around each one and cutting some 400 yards of new channel through the tailings. This added about 25 percent to the habitat for trout and also added greatly to the depth of the water. Log dams were installed in the new diverting channels to break up bare shallow stretches by forming small pools. Improvements began on June 7 and were completed on July 16, 1949. The next step contemplated was to seed the plant by air to provide more shade and cover to the stream.

Obviously, this was only a minor advance. Whether any further improvements were completed at that time is not clear. Andriano's report was probably submitted to officials of the Idaho Fish and Game Department. It outlined the work that had done and ended with a bill for $1,493.75, of which $932.25 was for the lease of a D-8 Caterpillar and operator from the Baumhoff-Marshall Company at a bargain rate.[27] One thing is certain: Andriano knew

FIGURE 16.1. The Idaho-Canadian Dredging Company's dredge was the last to dig in Boise Basin. Like the others, it left a lasting impression on the environment. Courtesy, Idaho State Historical Society, Boise, ISHS 72-112.b.

what he was about. Brief though it was, his approach was precisely what his successors would utilize wherever dredged-over watercourses were eventually restored elsewhere in the West.

Apparently, there was a long lapse before renovation was again seriously considered. When it came, it focused on the Idaho City area, where Mores Creek and its tributaries, Grimes Creek and Elk Creek, had all been heavily dredged, with at least 400 acres of tailings left in the vicinity. All open-valley streams had been left with highly disconnected, disrupted, and confined channels, waters classified by the US Environmental Protection Agency as "impaired" because they were too hot. In addition, there was almost no riparian vegetation and few or no floodplains in the wider valleys. In 1998 a project was organized by the acting ranger for the Idaho City Ranger Station, who approved one mile of Mores Creek for riparian restoration. Some minor stream-bank stabilization occurred, but no major work was done because of a lack of funds. What was accomplished was that over the next few years a group of individuals and private, state, and federal agencies was organized to support such a project.[28]

The "partners," or "stakeholders," as they were called, represented a wide range of interested contributors: the US Forest Service, the US Bureau of Land Management, the US Fish and Wildlife Service, the Idaho Conservation League, Boise County, the town of Idaho City, and a number of local landowners and citizens. Dredge magnate Fred Baumhoff's two sons, Bruce and Oscar—both of whom owned property in the basin—took an active part in the project. Later, Trout Unlimited joined as a partner, with a person based in Idaho City to coordinate and assist. Trout Unlimited had been in existence for over half a century. A national organization with more than 140,000 volunteers in 400 chapters around the country, plus a large staff of lawyers, scientists, and policy experts, it was dedicated to one simple cause: "To conserve, protect, and restore North America's coldwater fisheries and their watersheds." Among the appointee's other duties were to help define restoration based on the condition of the stream and to work with technical experts, private landowners, and federal, state, and local agencies to help design and implement projects. The person was also charged with developing educational activities that might be pertinent for both the project and the schools, as well as devising a media plan to work with Trout Unlimited's publicity staff in fund raising.[29]

The project aimed to restore nine miles of Mores Creek, seventeen miles of Grimes Creek, and three miles of Elk Creek—all on government land. The partners had developed a three- to five-year agenda divided into four phases, the first of which was for demonstration purposes—a showcase for local landowners to encourage their support of further restoration. Some preliminary work may have been done in 2005, but phase I really began its rehabilitation of half a mile of Mores Creek in the spring of 2006. After heavy equipment operators had pushed back the tailings thirty feet or so to create a floodplain, new riparian vegetation was planted, and a series of five rocks and stumps was installed to give "cultural diversity." Already, 4,000 donated sedges and rushes had been planted, along with nearly 2,000 willow saplings and 140 clumps of willows. During the summer and fall, 1,000 more trees, including cottonwoods and dogwoods, were put in by volunteers. Pam Smolczynki, Trout Unlimited's coordinator on the scene, appealed in a blog: "Getting your hands dirty . . . Want to plant a tree for trout? I could really use some hands to plant some beautiful cottonwood trees along the banks of Mores Creek, near Idaho City!"[30]

By this time in phase I, about 12,500 cubic yards of tailings had been moved from the stream's edge, at a cost of $228,000 for earth moving and structure installation. The engineering contract was $140,600. The sources of funding so far included the Southwestern Resource Advisory Committee, $227,500; the US Forest Service, $117,500; the National Fish and Wildlife Foundation, $85,000; the Idaho Conservation League, $20,000; US Fish and Wildlife Service, $12,500; and Trout Unlimited, $6,000. Nearly a dozen groups made in-kind donations—among them Idaho City High School, Boise County, the US Geological Survey, and countless volunteers from many groups. In this respect, Trout Unlimited stood out when it provided the salary of a water-quality analyst for three years with funds from the Tiffany and Company Foundation. It also did yeoman's work in publicizing the project and using it as learning tool.[31]

Phase I was expected to be completed in 2007, but phase II, to restore another half a mile of stream also on US Forest Service property, had to be rescheduled for 2008. With construction along Elk Creek and private lands downstream on Mores Creek, it was deferred even longer. Financing became more difficult because of the recession, according to Trout Unlimited: the organization was literally "cobbling together" what it could. Trout Unlimited added a $5,000 Embrace-a-Stream grant of its own through its Snake River branch, the US Fish and Wildlife Service added $136,000, while the Idaho Office of Species Conservation contributed $70,000. The work accomplished was more of the same: removing tailings to create more floodplains, planting vegetation to provide shade and habitat for both bull trout and animals, and building woody debris structures and boulder complexes to enhance the aquatic habitat for game fish. What was added was restructuring the channel with more natural meanders.[32]

When it was clear that more sustained funding was required, in 2006 the group applied to the BPA for a grant of $2,741,500. The sponsor was West Central Highlands Resource Conservation and Development, a grassroots regional agency committed to help rural communities better manage their resources. Its role seems to have been lending its name to the Mores Creek Project when funding was needed. The lengthy application was extremely detailed and indicated that the project was not as far along as it appeared. As well it might, the application focused on the intensive stream damage in Boise Basin and what the partners had already accomplished there: "The project involves all relevant local, state, and federal agencies, as well as the Shoshone-Piute Tribes. The diversity of groups and the early cooperation and collaboration improves the potential for meeting the variety of concerns and needs by the project. The project will offer an array of educational opportunities for the locals and provide restoration incentives for other applicable landowners." The application was quickly approved, but only with several important qualifications. Final sanction was contingent on a favorable step review by the ISRP, which asked that a number of issues in planning and implementation be clarified. After that, formal scrutiny by the Northwest Power and Conservation Council was essential. The council had already recommended a last-minute budget reduction, which lowered the funding for FY2007 from $1,042,400 to $665,000, that for FY2008 from $830,800 to $455,500, and that for FY2009 from $868,300 to $523,000. These decreases were based on the council's belief that not enough bull trout existed to justify the requested level, that its proposed funding was adequate to allow increased water-quality and resident fishery habitat improvement and still allow any remnant bull trout numbers to increase, and that it assumed that other cooperative funding efforts would continue.[33]

When state officials evaluated Mores Creek and its tributaries for sediment reduction in 2008, they noted significant progress in the restoration of Elk Creek. Slopes were being stabilized by placing water-spreading structures on topsoil when the first signs of erosion appeared—rock dams, logs, and logging slash dissipated overland flow and reduced sediment delivery in the creek. Streamside riparian buffers were used to filter storm-water pollutants. Officials also mentioned that the project was on schedule and were hopeful that it would be completed by early 2012, but additional funding was needed to finish the work. The next evaluation report, in 2009, observed that as part of the five-year project, construction of four large settling ponds had been completed to capture the runoff from the historic placer and hydraulic mining areas. During the spring rains, the ponds were put to the test and "functioned very well," and for the first time in many years the residents of Idaho City did not have to boil their drinking water. Both reports noted that the project had asked for a grant from the Bonneville Power Administration to complete the work, which meant that the project's funding had not been finalized.[34]

In the quest for rare-earth minerals, between 1956 and 1959 Porter Brothers Corporation dredged private land on upper Bear Valley Creek, an important salmon and steelhead spawning tributary of the Salmon River. In so doing, they obliterated 17,000 linear feet of the original channel of the stream and 10,000 feet of tributary channels. The creek channel not wiped out was de-watered for 45,000 linear feet, and the new artificial channels were not suitable for salmon habitat. In addition, the dredges left large amounts of unconsolidated overburden along 1¼ miles of the stream's floodplain. As a result, it was estimated that more than 17,657,333 cubic feet of fine sediment had entered the creek since the dredging began.[35]

In 1969 an attempt was made to correct part of the damage. The dredged section of Bear Valley Creek and the lower reaches of Casner Creek were diverted and channelized. This diversion failed several times, especially during a flood in 1984, which resulted in substantial downstream erosion and erosion of tailing materials into the stream. The Shoshone-Bannock Tribes then initiated a plan to reduce the amount of sediment entering the stream and enhance the creek as salmon habitat. Between 1985 and 1989, the high tailing banks were graded and vegetation—mainly grasses—was planted to create a new floodplain along 1½ miles of stream. Rehabilitation was completed in 1989 at a final cost of $2.8 million. The Bonneville Power Administration provided the funds, and the Shoshone-Bannock Tribes provided the manpower.[36]

There was another dimension to cleaning up after the Bear Valley dredges—the mill near Lowman in Boise National Forest. When the plant closed in 1960, Porter Brothers sold it to the Velsicol Chemical Company, but the title was soon transferred to the federal government. The mill site of 37 acres contained about 129,400 cubic yards of radioactive contaminated materials, including demolished building foundations and process structures, settling pond residues, and 28,000 cubic yards of vicinity property materials. The entire area was stabilized in a disposal cell in 1992 and licensed by the Nuclear Regulatory Commission two years later. These materials were compacted and covered with an engineered cap system that met Environmental Protection Agency standards for longevity, radon emanation control, and groundwater protection. The built-up cap consisted of a 1½-foot-thick layer of compacted clay soil that served as the main barrier against emission of radon from the cell and infiltration of rainwater into the cell. A 6-inch-thick layer of sand and gravel covered the barrier and served as the bedding layer for the overlying erosion-resistant cover. A 12-inch-thick layer of crushed basalt rip-rap material covering the top and side slopes provided protection from erosion by wind and surface water. A rock apron around the cell directed rainwater away from the cell structure and provided further erosion control. Responsibility for the cell was vested 90 percent in the US Department of Energy and 10 percent in the state of Idaho. An initial federal appropriation of $27,000 was made, with additional funding for future management support.[37]

Billions of dollars were poured into renovation of streams damaged by dredging, thanks to the federal government's passage of the Northwest Electric Power Planning and Conservation Act in 1980, as well as to the Bonneville Power Administration. The funding went mainly to areas that had national forests and Native American tribes. Not all watercourses impaired by dredging were so fortunate.

After dredging on the North Fork and Palouse Rivers, for example, Northwest Goldfields, Inc., left behind the usual tailing piles and wounded waterways. In 1999 the Five Star Restoration Program gave a grant of $10,000 for a cooperative effort among private landowners, boy scouts, community volunteers, a logging company, the Idaho Transportation Department, and the Palouse Clearwater Environmental Institute to restore riparian habitat along the North Fork of the Palouse River in an effort to enhance water quality. The partnership will stabilize and re-vegetate 1,000 linear feet of stream and establish a 100-foot buffer that will also be planted with native plants. The donor believed that involving the community in the project would provide hands-on education and awareness of the importance of riparian habitat and the imperative for restoration. Even so, this was a mere pittance as far as rehabilitating dredged ground was concerned. As the assessor for the Idaho Department of Environmental Quality noted in 2005, the boat only worked for a few years, "but the effects from that operation can still be seen today, especially on the lower miles of the North Fork Palouse."[38]

Prichard Creek fared no better after the Yukon Gold Company boat had its long run. There, in 1953 Muriel Sibell Wolle had seen miles of dredge tailings, some so high she could not see over them. A Spokane newspaperman in 1990 indicated that this was still true. "The tall, barren rock piles stand as a silent reminder of a mining era void of environmental concerns," he wrote.[39]

In 2004 a study sampled sites on Prichard Creek for streambed and suspended sediment containing lead, zinc, arsenic, and mercury. The study

FIGURE 16.2. Unrestored dredged-over ground left behind small seas of ugly tailings, as Bohannon Creek in Lemhi County demonstrates. Courtesy, Conley Collection, Lemhi County Historical Society and Museum, Salmon, ID.

found varying enrichment of these elements along the creek, but the presence of many earlier deep-level mines, placers, and mills on the stream made it difficult to isolate the contributions of the dredge.[40] There is no solid evidence that any real improvement followed. In other streams that went through the dredging ordeal, there is also a good deal of unfinished business. In Idaho the day of the huge, clanking monster machines that left behind an unsightly wake was over. Time would provide a poor successor—the recreational dredge, usually a small suction machine manned by a single operator. In Idaho as elsewhere, this dredge was restricted by state limitations.

NOTES

1. Idaho Fish and Game Department, Fisheries Management Plan, 2013–18, 155, fishandgame.idaho.gov/public/fish/planFisheries.pdf.

2. Nicholas Collias, "The Six Million-Dollar Watershed," May 24, 2006, n.p., http://www.boiseweekly.com/boise/ImageArchives?by=920180&oid=1024584&page=4.

3. Sarah Elizabeth Purdy, "Analysis of Dredge Tailing Pile Patterns: Applications for Historical Archaeological Research," MA thesis, Oregon State University, Corvallis, June 7, 2007, 81, 86, 88.

4. Phoebe Siddall, *South Fork Clearwater River Habitat Enhancement Nez Perce National Forest* (Portland, OR: National Government Publication, April 1992), 6, docs.streamnetlibrary.org/BPA_Fish_and_Wildlife/16475-2.pdf.

5. Northwest Power and Conservation Council, *Pacific Northwest Electric Power Planning and Conservation Act* (Act of December 5, 1980), 85, 100, http://www.nwcouncil.org/media/5227150/poweract.pdf; Northwest Power and Conservation Council, *Sixth Northwest Conservation and Electric Power Plan* (February 2010), 8, http://www.nwcouncil.org/energy/powerplan/6/plan.

6. Northwest Power and Conservation Council, *Sixth Northwest Conservation and Electric Power Plan.*

7. Phoebe Siddall, "Final Report," Executive Summary, 1; Kim Clarkin and Kathleen Moynan, *Red River Construction Project* (Elk City, ID: n.p., April 1992), 1.

8. Hudson Mann and Mark P. Von Lindern, "Crooked River, Idaho County, Idaho, 1897," Water Quality Status Report 80 (Boise: Department of Health and Welfare, Water Quality Bureau, 1988), 1, www.deq.idaho.gov/media/435458-wqs80_crooked_river_1988.pdf.

9. Siddall, "Final Report," Executive Summary, 1, 25.

10. *Federal Register* (Washington, DC) 77 (December 12, 2012): 73976–78.

11. Siddall, "Final Report," Executive Summary, 1; Clarkin and Moynan, *Red River Construction Project*, 3, 41.

12. Columbia Basin Fish and Wildlife Program: Projects and Portfolios, "Project 2002-072-00: Red River Watershed Restoration," n.p., https://www.cbfish.org/Project.mvc/Display/2002-072-00; Red River Stream Improvement Design, March 15, 2007, n.p., www.osti.gov/scitech/biblio/947608.

13. Lower Red River Meadow Restoration Project, *Biennial Report 1996–97* (July 2003), Executive Summary, 8–11, http://docs.streamnetlibrary.org/BPA_Fish_and_Wildlife/00004478-1.pdf.

14. Rolland R. Reid, "Placer Deposits of the Elk City Region," pamphlet 121 (Moscow: Idaho Bureau of Mines and Geology, July 1960), 8; "Newsome Creek Ecosystem Assessment at the Water Scale," Final Report (April 2002), 1-4, 4-4, www.fs.usda.gov/Internet/FSE_DOCUMENTS/stelprdb5353169.pdf.

15. "20086 Rehabilitate Newsome Creek—S. F. Clearwater River," part I, 3, 8, n.d., http://citeseerx.ist.psu.edu/viewdoc/download?doi=10.1.1.196.2129&rep=rep1&type=pdf; "Rehabilitate Newsome Creek Watershed, Annual Report CY 2008" (January 2009), 2, http://www.osti.gov/scitech/biblio/962414.

16. "Draft Environmental Impact Statement—Nez Perce Tribal Homepage," September 2005, 1–6, http://nezperce.org/Programs/data/NewsomeCreek-DEIS_8-29-05.pdf; "Rehabilitate Newsome Creek Watershed, FY 2010, Annual Report," ftp://ftp.nezperce.org/.../Watershed.../Newsome%20Annual%20Repo....

17. "20086 Rehabilitate Newsome Creek," part I, 5.

18. Columbia Basin Fish and Wildlife Program: Projects and Portfolios (under Portfolios and Projects/Portfolios), "2013 Geographic Review," n.p., https://www.cbfish.org/.

19. US Bureau of Reclamation, "Yankee Fork Tributary Assessment," in *Reclamation: Managing Water in the West* (Boise, January 2–12, 2012), 1, 3, 55, http://www.usbr.gov/pn/fcrps/habitat/projects/uppersalmon/reports/uppersalmon/yfta/YF%20Tributary%20Assessment%20Final%20Report.pdf; US Bureau of Reclamation, *Yankee Fork Habitat Improvement Project: Pond Series 2*, Draft Report (Boise, December 2011), 1-1, https://www.nwcouncil.org/media/accord200205900/Attachments/Attachment%20E/YankeeFork_DraftPrelimBDR_PS3_122311_Report.pdf.

20. Salmon River Habitat Enhancement, docs.streamnetlibrary.org/BPA_Fish_and_Wildlife/14383-5.pdf.

21. Colombia Basin Fish and Wildlife Program: Projects and Portfolios, *Project 2002-059-00 Yankee Fork Salmon Restoration*, n.p., https://www.cbfish.org/Project.mvc/Display/2002-059-00.

22. *Challis (ID) Messenger*, August 23, 2012.

23. *Project Proposal Request for FY 2007–FY 2009 Funding, Proposal 200205900, Project: Yankee Fork Salmon River Dredge Tailings Restoration Project*, n.p., http://cfw.nwcouncil.org/solicitation/components/forms/Proposal.cfm?PropID=516; Evelyn Galloway, *Yankee Fork Floodplain Restoration Project Implementation Plan 2008–2024: Project # 200205900–Yankee Fork Salmon River Dredge Tailings Restoration Project* (Fort Hall, ID, May 2008), n.p., https://www.nwcouncil.org/media/accord200205900/Attachments/Attachment%20A/2008%20Yankee%20Fork%20Floodplain%20Implementation%20Plan%202008-2024.pdf; Independent Scientific Review Panel, *Review of the Yankee Floodplain Restoration Project Implementation Plan for 2008–2018* (Portland, OR: Independent Scientific Review Panel, September 21, 2008), 1, http://www.nwcouncil.org/fw/isrp/isrp2008-11; US Bureau of Reclamation, *Reclamation: Managing Water in the West; Yankee Fork Tributary Assessment Upper Salmon Subbasin* (Boise: US Bureau of Reclamation, January 2012), 3, http://www.usbr.gov/pn/fcrps/habitat/projects/uppersalmon/reports/uppersalmon/yfta/YF%20Tributary%20Assessment%20Final%20Report.pdf.

24. Mark Fritsch to Council Members, "Decision Memorandum, Northwest Power and Conservation Council," July 26, 2012, http://www.nwcouncil.org/news/meetings/2012/08.

25. Bonneville Power Administration, Division of Fish and Wildlife, "Fiscal Year Budget(s) for Project 2002-059-00" (under Reports and Documents), in *Project 2002-059-00: Yankee Fork Salmon River Restoration*, n.p., https://www.cbfish.org/Project.mvc/Display/2002-059-00; *The Columbia Basin Fish and Wildlife News Bulletin* (Bend, OR, September 14, 2012), n.p.; Minutes of the Northwest Power and Conservation Council (Spokane, August 7–8, 2012), www.nwcouncil.org/news/meetings/2012/08/.

26. Josh Bernard, *The History of the Idaho Chapter of the American Fisheries Society* (Boise: Boise State University History Department, May 11, 2007), 3; "*J.J.N.P. Company v. State of Utah*," *Pacific Reporter* 655, 2nd series (1982): 1133; *Deseret News* (Salt Lake City), October 23, 2001.

27. Donald Andriano, "Report on the Grimes Creek Improvement Area," July 1949, 1–3.

28. Collias, "Six Million-Dollar Watershed"; Columbia Basin Fish and Wildlife Authority, "Narrative for Proposal 20072100," January 10, 2006, http://cfw.nwcouncil.org/solicitation/components/forms/ProposalOld.cfm?PropID=591.

29. West Central Highlands RC&D, updated December 29, 2015, http://www.idahorcd.org/wchrcd/; Boise National Forest (BNS), "Briefing Paper: Mores Creek Mine Site Watershed Restoration Project" (June 1, 2006), 1–2; Trout Unlimited, http://www.tu.org/connect/groups/idaho-trout-unlimited.

30. West Central Highlands RC&D, http://www.idahorcd.org/wchrcd/; BNS, "Briefing Paper," 2–3; Ted Trueblood Chapter, "Mores Creek Tree Planting," n.d., http://www.tedtruebloodtu.org/morescreek.htm.

31. Mores Creek Floodplain Project—Idaho RC&D. Apart from the cost of moving tailings and the engineering contrast, these figures undoubtedly combine both cash and in-kind contributions.

32. *Idaho Trout* (Spring 2008), 1, tedtruebloodtu.org/News/IdahoTroutNewsletterfinal.pdf; US Fish and Wildlife Service, "2007–2008 Conservation Funding Accomplishments, Snake River Fish and Wildlife Office," www.fws.gov/idaho/publications/RecoveryFunds08.pdf (no longer available); Columbia Basin Fish and Wildlife Authority, "Narrative for Proposal 200721000," January 10, 2006, http://cfw.nwcouncil.org/solicitation/components/forms/ProposalOld.cfm?PropID=591; BNS, "Briefing Paper," 4.

33. West Central Highlands RC&D, http://www.idahorcd.org/wchrcd/; Columbia Basin Fish and Wildlife Authority, "Narrative for Proposal 200721000"; "200721000—Mores Creek Watershed Floodplain and Habitat Restoration: Design and Implementation," 3, http://www.nwcouncil.org/fw/reviews/2007/reviews_detail.asp?id=591%2F; Administration and Budgeting—Columbia Basin Fish and Wildlife Authority, January 10, 2006, https://www.cbfish.org.

34. Idaho Department of Environmental Quality, "2008 Performance and Progress Report," 38, https://www.deq.idaho.gov/media/458878-performance_progress_report_2008_entire.pdf; Idaho Department of Environmental Quality, "2009 Performance and Progress Report," 27, https://www.deq.idaho.gov/media/458875-performance_progress_report_2009_entire.pdf.

35. *Bear Valley Creek 4b Justification* (Boise: Idaho Department of Environmental Quality and Lowman Ranger District, Boise National Forest, February 2011), 28, 40, https://www.deq.idaho.gov/media/463262-bear-valley-4b-justification-0211.pdf.

36. Ibid.; "Salmon River Habitat Enhancement Annual Report 1990" (Fort Hall, ID: Shoshone-Bannock Tribes, December 1991), 1–2, docs.streamnetlibrary.org/BPA_Fish_and_Wildlife/14383-5.pdf.

37. US Department of Energy, "Lowman, Idaho Disposal Site Fact Sheet," n.d., n.p., http://www.lm.doe.gov/Lowman/Documents.aspx#fs; US Department of Energy, Office of Legacy Management, "Lowman, Disposal Site" (December 12, 2011), http://www.lm.doe.gov/lowman/Sites.aspx.

38. Idaho Department of Environmental Quality, *Palouse River Tributaries Subbasin Assessment and TMDL* (January 2005), 28, https://www.epa.gov/.../tmdldocs/11288_Palouse%20River%20TMDL.pdf.

39. US Environmental Protection Agency, "Projects Funded by Five Star Restoration Program in FY99," water.epa.gov/ › ... › Wetlands › 5 Star Wetland Restoration and Grants; Muriel Sibell Wolle, *The Bonanza Trail* (Bloomington: Indiana University Press, 1953), 253; *Spokane Chronicle*, November 2, 1990.

40. Stephen E. Box, John C. Wallis, Paul H. Briggs, and Zoe Ann Brown, "Stream-Sediment Geochemistry in Mining-Impacted Streams: Prichard, Eagle, and Beaver Creeks, Northern Coeur d'Alene Mining District, Northern Idaho," *Scientific Investigations Report 2004-5284* (US Geological Survey, 2005), 18, pubs.usgs.gov/sir/2004/5284/sir2004-5284.pdf.

17

Overview

It had long been known that there was gold in the Snake River, and by the 1870s prospectors were able to mine the placers in scattered sites along the shore using sluices with burlap sacking for recovery. The main problem was how to capture the extremely fine-sized, elusive particles. In the belief that new technology could not only solve that difficulty but would be able to handle more ground profitably, officials of two Detroit companies put a pair of dredges to work on the Snake in 1882 near the modern site of Burley. This was the first gold dredging recorded in Idaho. The two suction machines were built under the patents of John F. Sanders of Ogden. The event proved to be an utter fiasco, perhaps even a fraud, but it soon touched off an ongoing boom of dredge building in the 1890s for use in the water and on the bars of the Snake that lasted well into the first decade of the twentieth century. Dozens of machines were constructed, but only one paid a dividend—a suction dredge on the same part of the river as the Sanders disaster.

Add to that another factor: the 1905 report of geologists David T. Day and Robert H. Richards on useful minerals in the black sands of the Pacific Slope,

DOI: 10.5876/9781607324751.c017

including Idaho. Commissioned by the US Congress, this was a major study of the heavy waste sands cast aside from sluice boxes in cleanups of a variety of types of placer mining, including dredging. When a host of experts analyzed these samples, the result showed gold in sizable amounts among the sixteen "salable" minerals in the black sands. Day called upon the mining profession to develop new machinery and new methods to break down the black sands more efficaciously and inexpensively. Idaho newspaper headlines went wild. One metallurgist called the Snake River the "largest and richest nature sluice box in the world, and perhaps the greatest deposit in the world."[1] "Process mania" set in, with sharp operators offering new and infallible formulas for handling the black sands. Some fraud was no doubt involved. Gold dredging in the Snake River was given new hope for a dozen years or so. The optimism generated went unrewarded: no viable solution was ever found for breaking down black sand or for dredging the fine gold of the river.

The Sweetser-Burroughs Mining Company was organized in 1893 for work on the Snake. Its story is worth discussing in some detail for several reasons. The company's suction dredge was the only one of the many boats that ran on the river to ever pay a dividend. This fact made it an advertisement, heralded in newspapers and mining journals, of what was actually possible, and it no doubt encouraged further dredge building. Moreover, it shows the actual working of an early boat on the river under very competent management. The company's bucket dredge was never successful, and it demonstrates the need in that situation for the owners to frequently move their machines to new locales. Since this was essentially a family operation, it also gives insight into life on the river.

Most of the gold dredged in Idaho came from the high placer districts: Pierce, Florence, Stanley Basin, Elk City, Salmon City, Warren, Boise Basin, Yankee Fork, and the South and Middle Forks of the Boise River. Such areas were similar in a number of ways. Gold was discovered in all of them in the early 1860s, they were all located in areas that had heavy winter weather and very poor transportation, and miners in all of them used the same standard equipment and methods in pre-dredge days. After a few years, they all had substantial numbers of Chinese miners, but there is no record of a Chinese person owning or operating a dredge in Idaho—unlike Sew Hoy in New Zealand, who owned at least four machines. Except for Yankee Fork and the two on branches of the Boise River, all the other districts were attempting to utilize dredges either in the late nineteenth or early twentieth centuries, although only the first boats in the Pierce district and Boise Basin had any measure of success.

The dredging history of the high placer districts might vary. In essence, Florence had almost none; the cream had been skimmed off in the first two years after discovery, and the few machines used afterward produced very little gold. The Stanley Basin was also a disappointment, first because of poor machinery, then because of difficult ground. The Warren district was a late bloomer. Early, a Marian steam shovel was used briefly, but it was not economical. A small Risdon worked for only two months in 1905 and was soon abandoned. Then came a long hiatus until 1931, when the Baumhoff and Fisher partners started up the first of two dredges in a lucrative operation that lasted eight years. Pierce, Elk City, and Salmon City all provided a mix of a few early failures and a good many more profitable operations, especially by small dredges.

Boise Basin was the richest and most enduring of the districts. If its history shows vividly the early dredging days, when companies went broke because of poor equipment and bad economic judgment, it also displays examples of small company success (the Moline Mining Company) and larger firms such as the Boston and Idaho Gold Dredging Company, which operated during all but two years between 1898 and 1916. Fred Baumhoff and two different partners ran the two boats for fifteen years, beginning in 1935. A number of productive runs by other boats were less spectacular but profitable.

Operations in some districts were performed by a single machine. Yukon Gold's venture on Prichard Creek was an example. Bringing a dredge 5,000 miles from the Klondike in 1917 by river, ocean, rail, and the last few miles by wagon must have been a challenging undertaking. Despite the loss of a year's time as the result of a sunken machine, the company still made money on its run. In 1922 the South Park Dredging Company started digging with a new Yuba near Featherville on the South Fork of the Boise River and worked steadily until spring of 1928, a highly profitable investment. On the Middle Fork of the same stream, in mid-June 1941 Boise King Placers began digging with a seven-and-a-half-foot Yuba and continued to October 1942, then sold its boat in what most certainly was a brief losing season.

Apart from the high placer areas, there were few successful dredge operations. Perhaps Northwest Goldfields, Inc., in the far northern Hoodoo district, was one: starting in 1940 in a run broken by the World War II ban, the company made an estimated $550,000. Another possibility is Dixie Placers in the Dixie district, with a dragline that completed six years of dredging.

The federal government normally kept a hands-off policy where gold dredging was concerned, but in several instances it did not. One of the latter cases with indirect consequences for dredging was the raising of the price of gold to $35 dollars an ounce in 1935. Amid the Great Depression, this move created a boom for both quartz and placer gold mining that lasted until World War II. Small dredges and a newcomer in the form of the dragline now became popular in Idaho and the rest of the West.

Another example was seen during the war, in 1942, when almost all gold mining was banned. Exempted were mining operations in which gold was a by-product of another mineral deemed strategic to the war effort. Only one Idaho dredge qualified, and it had to move to Montana to do so. A number of dredge towns were virtually deserted as workers enlisted or flocked to the Pacific Coast where the jobs were.

Still another official action that affected Idaho dredging occurred during the Cold War years, when the government sought to build stockpiles of strategic minerals, especially rare metals. Along with buying minerals abroad at high prices, part of the plan was to encourage subsidies to private enterprise to explore, discover, and develop these resources. Through several federal programs, at least 350 applications were filed in Idaho, of which only a small percentage were for rare or heavy metals that could be dredged. In the end, one of those that made the cut was the Ruby Meadows Mining Company, headed by George C. Kosmos, whose experience with the federal government took on aspects of a comic opera before it ended. Others were more successful. Porter Brothers Corporation eventually came up with a government contract to produce 525 tons of columbium-tantalum concentrates at Bear Valley, with a mill at Lowman. Meanwhile, Washington was doing its best to encourage private production of rare-earth minerals and succeeded in attracting two dredge outfits from the now idle Boise Basin. Both operated for several years on Big Creek in the Cascade Valley.

All types of dredges caused great damage to streams, whether actually working on them or beside them. The big machines modified channel courses, destroyed riparian vegetation by leaving windrows of tailings behind, and ruined the habitat for the many fish that inhabited the waterways. For the average Idahoan, this was the normal price for developing resources and creating jobs. The environment was not part of the equation. Then, beginning in the 1930s attitudes began to change: various interest groups in the state began to clamor for a halt to the despoiling of streams by gold dredging. The movement was slow, fought by the Idaho Mining Association, but with a ballot initiative in 1954, the reformers were able to put teeth in a watered-down version of a bill passed by the legislature the previous year. Unfortunately, the new law came after most of the dredging in Idaho had been completed. The damage had already been done.

It took another generation to show an interest in rehabilitating the streams, and when it occurred it was primarily a by-product of an action of the US Congress in 1980, when that body passed the Pacific Northwest Electric Power Planning and Conservation Act. Part of this law required the Bonneville Power Administration to spend part of its profits to mitigate the negative effects of the Columbia River hydroelectric system on fish and wildlife in Idaho, Oregon, Washington, and Montana. Since drainage of some of the Idaho rivers was on the Nez Perce Tribal Reservation, Native Americans would play a substantial role in the actual work of reconstruction, while Bonneville Power provided most of the funding—although there was often cooperation with the US Forest Service, the US Bureau of Reclamation, various Idaho state agencies, and private sources. Between 1984 and 2014, Bonneville Power spent billions of dollars on stream rehabilitation, much of it going to the Nez Perce and the Shoshone-Bannock Tribes. Included was a restoration plan for the South Fork Clearwater Sub-Basin, another for Yankee Fork, and still another for upper Bear Valley Creek. Boise Basin provides an exception. Its streams suffered more than any others, but restoration there was largely performed by a local group that raised money from many sources: individual volunteers, local landowners, high schools, and a wide variety of private, state, and federal agencies. An application for funding was made to Bonneville Power, but there is no record of the group having received it. Elsewhere, a number of dredged-over sites received little or no attention.

The story of gold dredging in Idaho or any other western state would not be complete without mention of the development of the California-type bucket-line dredge, which blended the New Zealand style with the first American machine and was gradually improved over the years. Western dredge men came to consider it the best in the world. It was not the product of any one manufacturer—the name became a kind of generic term. Whether it was a Yuba made in California, a Bucyrus built in Wisconsin, or a Marion constructed in Ohio, it was still a California-type dredge and considered to be unsurpassed in its field.

Note

1. George W. McGhee, "The Black Sands of the Snake River, Idaho," *Ores and Metals* (Denver, CO) 14 (August 20, 1907): 313.

Epilogue

Serious scholars in general have tended to ignore the dredge mining industry in the United States. Lacking are comprehensive treatments of the subject. I was once naive enough to believe I could write a complete history of gold dredging in the West, but eventually, after publishing a book on Alaska dredging, I was overwhelmed by the mass of material in libraries and on the Internet. California, Colorado, Montana, Oregon, and Idaho were all major producers of dredged gold, and they all invited the historian's attention. I chose Idaho for the current book. Why? In part because I grew up in the southern part of the state, just a stone's throw from the Snake River, and I was intrigued by the many early unsuccessful efforts to dredge the river's fine gold. At least thirty-seven dredges of one sort or another were built for that purpose. One can marvel at how all the newspaper and mining journals' hype regarding a single dredge dividend, combined with the black sand promises, kept miners' hopes high for more than two decades. The gold was there, but there was no means of extracting anything but a small fraction of it. In 1985 historians at the Idaho State Historical Society estimated that more

DOI: 10.5876/9781607324751.c018

than $2 million worth of gold had been taken out of the bars along the Snake by small-scale miners using more primitive equipment.

In addition, Idaho was a major producer of dredge gold and offered much placer ground for the industry in diverse parts of the state. Over the years, Idaho usually ranked fourth among US states in gold production. Then, too, Idaho seemed more manageable in terms of available material, although it was certainly less significant than California.

Gold dredging was obviously of considerable importance to the Idaho economy. Most of the companies were organized in other states, with much capital raised elsewhere. Success meant that profits flowed elsewhere in such cases, but local investors, labor, and Idaho tax receipts benefited. Work crews on an operating bucket dredge were sizable because of the need for three shifts in hopes of running the maximum twenty-four hours per day, seven days a week. Most of the machinery came from out-of-state manufacturers, which meant a further cash drain, although it is surprising how many times some dredges were rebuilt.

There was far less litigation in dredge mining than resulted from hard-rock mining, where the courses of lodes were often disputed. In Idaho, dredge mining placer ground was usually purchased or leased, and the bounds were clearly defined. There were a few cases stemming from whether property was farmland or mineral land, a few more involving injuries or deaths of employees, and the classic case of the Idaho Gold Dredging Corporation suing the Boise Payette Lumber Company over the presence of oil and grease on ground it had purchased. In general, though, "courtroom mining" can be said to have been confined to lode mining.

Any dredge operator faced many challenges. To begin with, he had to contend with the elements. The rich high-country placers were located in hard-to-reach areas with extremely poor transportation. To reach a place like Florence from the railhead meant traveling the last 110 miles or so over primitive roads and rough trails using pack horses and teams. It took eight teams to transport a boiler alone. The route to Elk City was no better: to move forty tons of machinery on a crude road that went over a 6,500 pass required thirteen teams.

Cold weather was another factor. In Alaska, where ground had to be thawed using a bristling forest of pipes that carried water down under pressure to percolate through the pay gravel, the working season was understandably short. The same was true of most high-country Idaho dredges, although in Boise Basin digging sometimes continued year-round.

Water has always been the key to most placer mining, all the more so with dredge mining. In dredging, water was essential for floating the boat and for pumps to scrub the grizzly and wash the gold. The addition of electric motors to power the digging mechanism, the grizzly, the tailings stacker, one or two heavy spuds, and perhaps several other winches sometimes meant that a company had to build its own powerhouse by bringing the water eight or ten miles through rough terrain before dropping it down under pressure to the generator.

In the early days of dredging, too many investors experienced a short run of the machine, then closure and bankruptcy. That situation changed when gold dredging came of age with advancing technology and knowledge. A good, experienced engineer could drill test holes in a pattern on placer ground and predict very closely what the production would be. Gold dredging now had less risk than any other form of mining. The California-type dredge was an excellent machine and the key to successful dredging, but it was not infallible. No engineer could anticipate what other pitfalls might occur: the breakdown of equipment, the sinking of the dredge, or the dredge catching fire and burning. In the Stanley Basin a layer of talcy gravel that contained "rusty" gold stymied dredge men because in this "robber clay" the gold would not amalgamate. Large rocky formations might require dynamite to clear the way, and long stretches of unproductive ground to work across were both expensive and time-consuming. Finally, there was the ultimate obstruction: running out of ground, which meant either enduring the grueling task of moving the dredge or giving up.

The dredges were responsible for working low-grade ground that would never have been touched otherwise, and they were sometimes able to dig and make a profit from the leavings of hydraulicking. Idaho's placer gold was relatively close to the surface, thus dredges there called for digging ladders to go down only 20 feet or 35 feet at the most, whereas many California dredges were able to dig much deeper. A number of the latter were able to reach 50 feet or 60 feet, and a few were able to go down 100 feet or even 150 feet.

California, the Klondike, and Alaska were noted for large companies operating multiple fleets of gold dredges. In the north, the Fairbanks Exploration Company worked a total of eight bucket dredges; in California, while a number of small operators produced considerable gold, most of the richest fields involved three large concerns—Oroville Dredging Company, which was British-financed; Natomas Consolidated, which drew its capital from San Francisco; and Wendell Hammon's Yuba Consolidated Gold Fields, a Boston firm—each with a small armada of dredges. Hammon controlled twenty-two dredges over a period of sixty-five years. This was not generally so in Idaho. True, the Yukon Dredging Company, which dredged in the Coeur d'Alenes and was owned by the Guggenheims, had a number of dredges in Alaska, Canada, and California, but it had only one in Idaho. Most operators in Idaho owned a single dredge; several ran two or three, and one partnership ran four.

Brief Bibliographic Essay

Sources for this study are varied and widely scattered. Primary materials are found in a number of libraries and include the following collections: the George P. Hurst MSS in the Stanford University Archives, Palo Alto, CA; the rich papers of Charles Romanowitz in the Rasmuson Library, University of Alaska, Fairbanks; the Fred L. Morris MSS in the Bancroft Library, University of California, Berkeley; the Charles Janin Papers in the Huntington Library, San Marino, CA; the Paul E. Tracy MSS in Special Collections and Archives, Albertsons Library, Boise State University, Boise, Idaho; the Yuba Manufacturing Company Papers in the Meriam Library, Chico State University, Chico, California; and the Jonathan Bourne Jr. MSS in the University of Oregon Library, Eugene.

At the same time, a wealth of primary sources is available on the Internet. Both chapters 16 and 17 draw heavily on such materials, with access to all kinds of documents, including much correspondence. I have utilized Google, especially Google Patents. The countless newspaper clippings in the Kimble Northwest History Database in the Washington State University Libraries

DOI: 10.5876/9781607324751.c019

Digital Collections have been invaluable in this study. So, too, has been Utah Digital Newspapers on the Internet. Every one of the many citations from Utah newspapers and journals comes from that source.

Technical publications have also been important, especially *Mining and Scientific Press*, *Engineering and Mining Journal*, *Mining World*, *Mining and Engineering World*, and *Transactions of the American Institute of Mining Engineers*. Widespread use has also been made of state pamphlets, circulars, and professional papers dealing with specific Idaho mining districts, among them many issued by the Idaho Bureau of Mines and Geology, the University of Idaho, and the Idaho State Historical Society Reference Series. Likewise, the yearly report of the state inspector of mines provided much information. Many documents issued by the federal government proved pertinent. For example, *Mineral Resources* was published annually by the US Geological Survey from 1882 through 1923 and by the US Bureau of Mines from 1924 through 1931, followed each year with the *Minerals Yearbook*, which reviews the mineral and material industries of both domestic and foreign countries.

Non-technical references include the *Idaho Reports*, for a number of court cases, and the *Idaho Daily Statesman* (Boise), both paper and digital. Gold dredging was always news in the latter. Also deserving mention is Merle W. Wells, *Gold Camps and Silver Cities*, Bulletin 22 (Moscow: Idaho Department of Lands Bureau of Mines and Geology, 2nd edition, 1983). Two studies by Sharon A. Murray are worth noting: "The Mount Vernon Dredge," typescript MSS (1985 or 1986), copy on file at the Nez Perce National Forest Supervisor's Office, Grangeville, Idaho, and "A Bucket Full: Boise Basin's Dredging Heritage, 1898–1951," *Idaho Yesterdays* 34 (Fall 1990).

Several MA theses cast light on Idaho gold dredging. One is Cletus R. Edmundson's "A History of Warren, Idaho: Mining, Race, and Environment" (MA thesis, Boise State University, August 2012). Another is Darin W. Vrem's "A History of Idaho's Dredge Mining Industry and the Dredge Mining Control Movement 1896–1970" (MA thesis, Boise State University, November 2004). A third is Sarah Elizabeth Purdy's "Analysis of Dredge Tailing Patterns: Applications for Historical Archaeological Research" (MA thesis, Oregon State University, Corvallis, June 2007).

Index

Page numbers in italics indicate illustrations.